Mechanics of Rotor Spinning Machines

Mechanics of Rotor Spinning Machines

By
Ibrahim A. Elhawary
Alexandria University, Alexandria, Egypt

CRC Press
Taylor & Francis Group
Boca Raton London New York

CRC Press is an imprint of the
Taylor & Francis Group, an **informa** business

CRC Press
Taylor & Francis Group
6000 Broken Sound Parkway NW, Suite 300
Boca Raton, FL 33487-2742

First issued in paperback 2020

ISBN-13: 978-0-367-57234-1 (pbk)
ISBN-13: 978-1-4987-1664-2 (hbk)

Library of Congress Cataloging-in-Publication Data

Names: El-Hawary, M. E., author.
Title: Mechanics of rotor spinning machines / Ibrahim Abdou El Hawary.
Description: Boca Raton : CRC Press, Taylor & Francis Group, 2018. | Includes bibliographical references.
Identifiers: LCCN 2017025814| ISBN 9781498716642 (hardback : acid-free paper) | ISBN 9781315371122 (ebook)
Subjects: LCSH: Spinning machinery. | Rotors--Vibration.
Classification: LCC TS1483 .E42 2018 | DDC 677/.02852--dc23
LC record available at https://lccn.loc.gov/2017025814

Visit the Taylor & Francis Web site at
http://www.taylorandfrancis.com

and the CRC Press Web site at
http://www.crcpress.com

I dedicate this work

to

my family and my grandchildren:

Hossam Eldin, Hana, and Saleh

Contents

Preface

There is a familiar statement made by both textile technologists and designers of textile machines: "The textile technological requirements are the basic resources for the textile machine designer." For example, the rotor spinning machine has three technological functions: (1) the sliver opening; (2) twisting, that is, inserting a twist in the strand of fibers that collect on the rotor's collecting surface during their withdrawal via the doffing tube; and (3) packaging, that is, winding the yarn on a spool at the top of the machine. Therefore, the designer has established three mechanical items: the opening device (feed roll and opening roll), the twisting mechanism (rotor), and the winding head. The running mechanical performance of these mechanisms will reflect the quality of the produced rotor-spun yarn, and maintaining the optimum mechanical conditions of the working element of the mechanism via various types of maintenance will be an additional factor for achieving a competitive yarn quality/price ratio in the global market. Knowledge of the previously mentioned terms and systems can be attained through the study of the mechanics—kinematic or dynamic—of the rotor spinning machine's elements or mechanisms. This textbook will help students of engineering institutions, engineering graduates, and the textile industry. Also, this textbook will help in prolonging a spinning machine's lifespan and improving the production of quality rotor-spun yarn.

The interaction between the machine and the processed material is responsible for 40% of product quality. We will produce a series of handbooks and textbooks on topics in the field of textile machine mechanics such as rotor dynamics (rotor spinning machines), ring-spindle vibrations (ring spinning machines), the vibrations of rotors, spindles and drafting systems, stresses in textile rotating masses, and so on. It must be highlighted that the student, engineer, and reader of such a series must have a good general background in mechanical engineering and mechanical vibrations especially.

Acknowledgments

The following scientists and institutions provided scientific materials and reviews based on their knowledge and teaching experiences that greatly helped us in preparing the current and future editions and textbooks.

Alexandria Ivanovich Makarov and Erick Alexvich Popover
Department of Technological Machines and Mechatronic Systems
Moscow State University of Design & Technology, Moscow, Russian Federation

Hosny Aly Soliman and Wael A. Hashima
Alexandria University, Alexandria, Egypt

About the Author

Ibrahim A. Elhawary is a professor emeritus at the Textile Engineering Department (TED) at Alexandria University and a registered professional engineer (RPE) with the state of Alexandria (General Engineer Syndicate [GES]). He obtained his PhD in mechanical vibration of heavy continuous filament twisting ring spindles from Moscow State University of Design & Technology in 1978. He earned two diplomas of engineering (BSc and MSc) in textile engineering from Alexandria University, Egypt, in July 1967 and 1972, respectively.

His professional experiences are teaching and research assistant from 1967 to 1973, PhD from 1973 to 1978, and a professor of textiles since 1991 to date.

Dr. Ibrahim has contributed to writing two reference books, namely, *Advances in Yarn Spinning Technology* (Woodhead, 2010) and *Textiles and Fashion* (Woodhead, 2014). Dr. Ibrahim has acted as one of the co-editors for *Cotton Fiber to Yarn Manufacturing Technology* (Cotton Inc., 2001) and *The Spinning Textile Magazine*, India. He is the author of several scientific articles, two text books for Alexandria University and Mansoura University, Egypt, and one reference book for the Egyptian Textile Industry. Dr. Ibrahim has been a visiting professor at Auburn University, Alabama; VJTI, Mumbai, India; Baroda University, Mumbai, India; JTI Delhi, India; and Bahir Dar University, Ethiopia; and has participated in several international conferences.

1

Rotor Dynamics

1.1 Introduction

Figure 1.1a shows the assembly of a rotor unit. The assembly consists of

1. A spindle (short shaft) with a diameter of φ10 mm that carries the rotor at its left end, while at its right end it carries a wharve (driving pulley) for the shaft (spindle). Between the rotor and the wharve, there are two bearings (antifriction type): one near the rotor and the other near the wharve; these two antifriction bearings are the carriers for the spindle.

2. A hollow-cup rotor (turbine/pot/camera with a maximum diameter φ69 mm) where the transferred fibers (via a transport duct or tube) from the opening device are deposited onto the rotor's collecting surface in layers (back doubling) to be withdrawn and twisted simultaneously to form the rotor-spun yarn. The rotor is the twisting element of the rotor spinning machine where its revolutions per minute (RPM) can reach up to 150 kRPM ($k = [E + 3]$). The base of the rotor is channeled or finned to increase its surface area to act as a cooler for the rotor body, where the generated heat at the spindle's bearings can heat the rotor body.

3. The wharve is a hollow, cylindrical, rigid body that is permanently fixed to the right end of the spindle. It is the driver for the spindle and rotor and is driven itself by a tangential belt drive. The wharve diameter is φ18 mm.

4. The bearings of the rotor, which are of the antifriction type, will be discussed in detail in the next chapter.

5. The outer gland (sleeve) of the bearings. This envelops the bearings from outside and plays the role of the bearing outer race. The gland's outer diameter is φ22 mm. The sleeve fixes the rotor assembly, which is set to the rotor spinning machine's chassis. The gland's length is 56 mm.

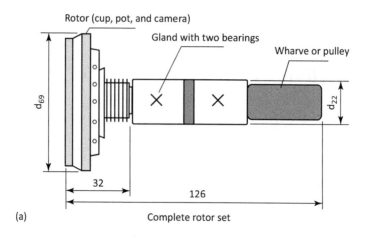

(a)   Complete rotor set

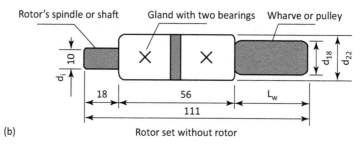

(b)   Rotor set without rotor

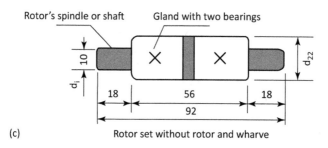

(c)   Rotor set without rotor and wharve

FIGURE 1.1
(a–c) Exploding view of a rotor set.

Figure 1.1b shows the rotor set without the cup where the spindle on the left end is naked. The rotor length of the set is 111 mm: the naked left spindle end is 18 mm, the gland length is 56 mm, and the wharve length is 37 mm.

Figure 1.1c shows the rotor set without either the cup or the wharve where the two spindle ends are naked. The total length of the set is 92 mm: 36 mm for the naked spindle's end and 56 mm for the gland's length.

1.2 Rotor Vibrations

The dynamics of the open-end (rotor) spinning machine's rotor refer to three items: (1) the critical speed of the rotor; (2) the amplitude of the mechanical vibrations; and (3) the dynamic reactions in the rotor's bearings.

1.2.1 Rotor's Critical Speed

The critical speed is the speed at which the rotating speed of the spindle with the rotor and wharve is equal to the circular natural frequency of the rotating systems. Another definition is the speed at which the amplitude of the mechanical vibrations is at its maximum. During the spinning process, the operator must avoid rotors running in the critical speed zone (resonance zone) due to the high mechanical disturbance of the machine and the poor quality of the yarn. Therefore, the designer and the engineers must select a safe region for rotor operation as shown by Equation 1.1 and the graph in Figure 1.2.

The equation is

$$0.7N_{cr_1} \geq N_w \geq 1.4N_{cr_2} \tag{1.1}$$

where:

$N_{cr_{1,2}}$ = Critical RPM of the rotation masses (1st and 2nd, respectively)
N_w = Working speed of the rotating masses (spindle, cup, and wharves)

Figure 1.2 shows the change in the amplitude of the vibrations with the changes in the rotating speed, and indicates the safe working zone during the spinning process. This curve is called the *resonance curve*.

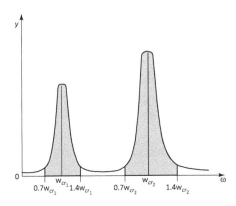

FIGURE 1.2
Amplitude graph for a blade with a mass distributed uniformly across its length (resonance curve). y = Amplitude of vibration; ω = rotor or wharve radian speed in seconds; ω_{cr_1} = first critical speed; ω_{cr_2} = second critical speed; shaded areas = safe running regions.

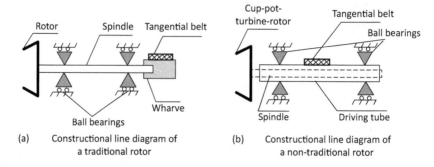

FIGURE 1.3
(a,b) Constructional line diagrams of a traditional and a non-traditional rotor.

Figure 1.3a and b shows the construction of the rotor set using line diagrams. Figure 1.3a shows a constructional line diagram for a conventional rotor set, including the cup (rotor), the spindle (shaft) with two bearings, and the wharve with a tangential belt drive. The shaft is an elastic element while the rotor and wharve are rigid bodies.

Figure 1.3b shows a constructional line diagram for a non-traditional rotor where the rotor's spindle is fitted inside a tube and plays the role of the wharve where the two bearings surround the driving tube (wharve substitution). Both of the constructional line diagrams will be used as a basis for calculating a rotor set's critical speed, from which we can determine the optimum work or running speed of the spinning process.

1.2.2 Influence Coefficients

Influence or effective coefficients play a significant role in determining the critical speed of the weightless (light) rotor's spindle. The definitions for these coefficients are (see Figure 1.4)

a_{11} = Displacement per unit load, that is, $m/N = m.N^{-1}$
α_{11} = Angular displacement per unit force, that is, $1/N = N^{-1}$,
b_{11} = Linear displacement per unit moment, that is, $m/m.N = N^{-1}$
$\therefore |\alpha_{11}| = |b_{11}|$
β_{11} = Angular displacement per unit moment, that is, $1/m.N = [m.N]^{-1}$

Furthermore, S is the spring constant (elastic coefficient), which is the force per unit displacement, that is, $N/m = [N.m^{-1}]$. Therefore, the multiplication of $a_{11} \times S = 1$ can be used for any elastic shaft irrespective of the bearing type (long bearing short shaft or shaft bending stiffness and the location of the bearing) (see Table 1.1).

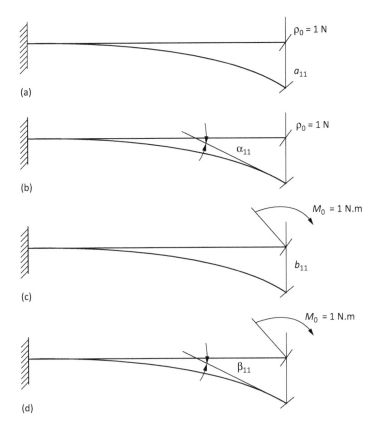

FIGURE 1.4
Influence coefficients. (a) Unit load a_{11}. (b) Unit load α_{11}. (c) Unit moment b_{11}. (d) Unit moment β_{11}.

It is important to note the following:
For calculating the influence coefficient, we apply the equation of the spindle elastic line or the deflection curve. Hence, we must use any technique from the strength of material science (SMS) such as double integration, Gastiliano, and so on. Then, the rotating speed in RPM can be calculated.

1.2.3 Equivalent Mass

When the rotor's spindle (shaft) transfers from massive to weightless, the equivalent mass is determined by the equality of the critical speed of the massive shaft with the critical speed of the weightless spindle; it has a different value due to the type of shaft bearings and their locations along the shaft length. In addition, the bending stiffness of the spindle plays a role. For example,

$$m_e = \frac{33}{140} \times M = 0.236 \times M \tag{1.2}$$

TABLE 1.1

Spring Constant and Critical Speeds of Different Calculation Schemes

Spring Constant (S)	Critical Speed ω_{er}	Calculation Schemes
$\dfrac{48EJ}{l^3}$	$\sqrt{\dfrac{48EJ}{l^3 m}}$	
$\dfrac{3EJ}{l^3}$	$\sqrt{\dfrac{3EJ}{l^3 m}}$	
$\dfrac{EJ}{0.093l^3}$	$\sqrt{\dfrac{EJ}{0.093l^3 m}}$	
$\dfrac{192EJ}{l^3}$	$\sqrt{\dfrac{192EJ}{l^3 m}}$	
$\dfrac{3EJ}{b^2 l}$	$\sqrt{\dfrac{3EJ}{b^2 lm}}$	

Source: Makarov A.I., 1968, *Construction & Design of Machines Used in Yarn Production,* Machines Design Press, Moscow, RFU.

where:

m_e = Equivalent mass of the rotor's shaft

M = Mass of the spindle $= \rho \times V$

In practice, the value of the equivalent mass is too small compared to the rotor's and the wharve's mass; therefore, when we calculate the critical speed of the rotor set, it can usually be neglected. But if the shaft's mass alone is massive, we can use its mass unit per length to calculate its first critical speed.

Usually, both the influence coefficient and the equivalent mass are used to calculate the first critical speed of the rotor set, as will be shown later.

1.2.4 Critical Speed of Non-Conventional Rotor

Figure 1.5 shows a scheme for calculating the critical speed of a non-traditional rotor. The rotor's spindle has two views: a straight-line view (static status) and a deflected view (dynamic status). In both cases, the spindle is completely fixed at its left end (long bearing), while its right end carries the rotor. The shaft is considered an elastic body while the rotor is a rigid body.

Legend for Figure 1.5

ℓ = Length of the static spindle (shaft) from its left to right end

Point (1) = A common point between the spindle (elastic element) and the rotor (rigid body)

s = Rotor's center of mass

k = Distance from point (1) and s

φ_1 = Angle of inclination of the elastic line's tangent at point (1)

Point $(1^\backprime)$ = New location of point (1) in the dynamic form

y_1 = Linear displacement of point (1)

$y_2(y_s)$ = Linear displacement of point (2) or $p(s)$ due to the dynamic form

p_1 = Inertia force of the equivalent mass at point (1)

$p_2(p_s)$ = Inertia force of the rotor mass

M_g = Rotor's gyroscopic moment (its direction is always to decrease the amplitude of vibration) where

$$M_g = \left(I_0 - I_q\right).\omega^7 \varphi_1 \tag{1.3}$$

I_0 = Rotor's mass moment of inertia about the axis of rotation or the polar mass moment of inertia of the rotor

I_q = The equatorial rotor's mass moment of inertia when the axis is perpendicular to the axis of rotation

ω = Radian speed of the rotor

φ_1 = As defined previously

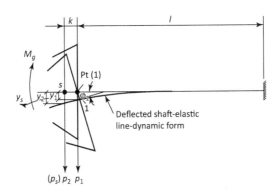

FIGURE 1.5
Schematic line diagram of a non-traditional rotor for calculating the critical speed.

By evaluating γ_1, φ_1, p_1, p_2, and M_g, and with the aid of the influence coefficients a_{11}, α_{11}, b_{11}, and β_{11} and the matrices technique, we can deduce the following equations for calculating the critical speed of the rotor:

$$\omega^4(BC - AD) + \omega^2(D - A) + 1 = 0 \qquad (1.4)$$

where:

$$\omega = \text{Radian speed of the rotor}$$
$$A, B, C, \text{ and } D = \text{Constants } A, B, C, \text{ and } D$$

$$
\left.
\begin{aligned}
A &= a_{11}(m_1 + m_R) + k\alpha_{11}.m_R \\
B &= \alpha_{11}(I_0 - I_q) - a_{11}.m_R.k - k^2 a_{11}.m_R \\
C &= b_{11}(m_1 + m_R) + k.\beta_{11}.m_R \\
D &= \beta_{11}(I_0 + I_q) - b_{11}.m_R.k - k^2\beta_{11}m_R
\end{aligned}
\right\} \qquad (1.5)
$$

where a_{11}, α_{11}, b_{11}, and β_{11} are the influence coefficients of the rotor's weightless shaft. For the non-traditional rotor's shaft:

$$
\left.
\begin{aligned}
a_{11} &= l^3/3EI \\
\alpha_{11} &= b_{11} = \ell^2/2EI \\
\beta_{11} &= \ell/EI
\end{aligned}
\right\} \qquad (1.6)
$$

where:

EI = Rotor's spindle bending stiffness

E = Young's modulus or modulus of elasticity in tension (compression) = 206 GPa

I = Shaft's inertia of cross section = $\pi d^4/64 = 0.05 d^4$

By solving Equation 1.4, we can find the first and second critical speeds where the first critical speed is due to the inertia force and the second critical speed is due to the gyroscopic moment. Accordingly, the first critical speed is relatively small while the second is relatively high. For the spinning rotor, in practice, we need the first critical speed only to calculate the rotor's working speed (running speed). In rotor spinning, the gyroscopic moment on the rotor is always decreasing the shaft's deflection; therefore, some heavy spindles are fitted at its upper end with dummy discs to decrease the amplitude of mechanical vibrations. Equation 1.4 can be applied to the calculation of the critical speeds for both the rotor (rigid body) and the wharve (rigid body) by changing the geometrical massive characteristics for each case.

1.2.5 Critical Speed of a Traditional Rotor

An exploded view of a traditional rotor is shown in Figure 1.6a through d.

Figure 1.6a shows a constructional scheme of a conventional rotor set (spindle, wharve, and rotor). Figure 1.6b shows the spindle (shaft) as a naked elastic body without a wharve or a rotor. The massive shaft of the spindle will be used for calculating its critical speed. Figure 1.6c shows the spindle with only the wharve fitted. The spindle here will be treated mechanically as weightless. Figure 1.6d shows the spindle as a weightless shaft with a fitted rotor.

Legend for Figure 1.6a through d

2ℓ = Total length of the spindle = $\ell_2 + 2\ell_1$

ℓ = Half spindle length = $\left(\dfrac{\ell_2}{2}\right) + \ell_1$

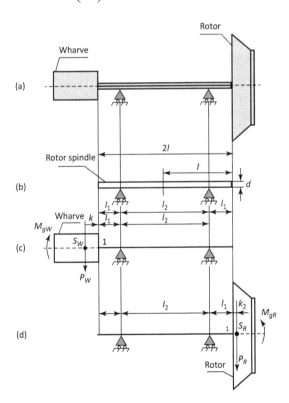

FIGURE 1.6
(a–d) Schematic exploded view of a traditional rotor for calculating the critical speed.

ℓ_1 = Free length of the spindle

ℓ_2 = Span of the spindle

d = Spindle's diameter

k_1 and k_2 = Distance from centers of gravity to point (1) for wharve and rotor, respectively

Figure 1.6a shows a complete rotor set (traditional) with a wharve, spindle, and rotor.

Figure 1.6b shows a spindle (shaft) only—a massive shaft. The spindle is considered an extended beam with two over-hangings. The shaft has two supports (bearings-roller type).

Figure 1.6c shows a rotor set without a rotor.

s_w = Wharve's center of gravity

k_1 = Distance from point (1) to s_w

Point (1) = Common point between the spindle and the wharve

p_w = Inertia force of the wharve

M_{gW} = Gyroscopic moment on the wharve—its direction is to decrease the amplitude of mechanical vibrations

ℓ_2 = Span length between point (1) and rotor center of gravity S_R

P_R = Inertia force on the rotor

M_{gR} = Gyroscopic moment of the rotor

S_W and S_R = Masses of the wharve and rotor, respectively

P_W and P_R = Inertia force of the wharve and rotor, respectively

The concept of the critical speed calculation is that it will be calculated individually for each figure before the resultant (total) critical speed of the rotor set is obtained using the Dynkerly formula:

$$\frac{1}{\omega_t^2} = \frac{1}{\omega_s^2} + \frac{1}{\omega_w^2} + \frac{1}{\omega_R^2} \tag{1.7}$$

where:

ω_t = Total radian critical speed of the rotor set

ω_s = Radian critical speed of the spindle (alone)

ω_w = Radian critical speed of the wharve (alone)

ω_R = Radian critical speed of the rotor (alone)

The total radian speed ω_t can be converted into a rotational speed in revolutions per minute using a well-known formula:

$$N_t = \omega_t \times 60 / 2\pi \cong 10\omega_t \tag{1.8}$$

1.2.6 Amplitude of Vibration

For the rotor of the rotor spinning machine, the amplitude of the mechanical vibrations is not important because of the small size and the dimensions of the spindle.

1.2.7 Dynamic Reactions

The dynamic unbalances in the rotor and the wharve create dynamic reactions in the shaft's ball bearings. The nature of these reactions will be studied in detail in Chapter 2.

1.3 Extra Note

1.3.1 Influence Coefficients

The influence coefficients $(a_{11}, \alpha_{11}, b_{11},$ and $\beta_{11})$ for a traditional rotor are

$$
\left.
\begin{aligned}
a_{11} &= \frac{\ell_1^2(\ell_1+\ell_2)}{3EI} \\[2mm]
\alpha_{11} = b_{11} &= \frac{l_1(3\ell_1+2\ell_2)}{6EI} \\[2mm]
\beta_{11} &= \frac{(3\ell_1+\ell_2)}{3EI}
\end{aligned}
\right\}
\tag{1.9}
$$

where:
EI = Spindle bending stiffness
E = Young's modulus = 206 GPa (steel)
I = Spindle cross-section area's inertia = $\pi.d4/64.(\pi\varphi^4/64)$

1.3.2 Critical Speeds

The critical speeds of the wharve and rotor as separate entities are calculated using Equation 1.4.

1.3.3 Naked Spindle

The critical speed of a naked spindle (alone) is calculated as a massive shaft by

$$
\omega_s = \frac{\alpha^2}{\ell^2}\sqrt{\frac{EI}{m}}
\tag{1.10}
$$

where:
- ω_s = Spindle critical radian speed as a massive shaft
- α = Critical speed constant = 2.22 (value taken from special tables)
- EI = Spindle bending rigidity
- E = Young's modulus = 206 GPa

- I = Cross-sectional area inertia of the shaft = $\dfrac{\pi d^4}{64} = 0.05d^4 \left(0.05\varphi^4\right)$
- m = Spindle's mass per unit length = πA
- ρ = Steel density (kg/m³)
- A = Spindle's cross-sectional area

1.3.4 Inertia Force

The inertia force on the equivalent mass of a massive spindle at point (1) is P_1:

$$P_1 = m_e.\omega^2.y_1 \tag{1.11}$$

where y_1 is the linear displacement of point (1).

The inertia force on the wharve and rotor, respectively, at their centers of gravity s_w and s_R are

$$P_2\left(P_s\right) = m_w.\omega^2.y_2\left(y_s\right) \tag{1.12}$$

and

$$P_2\left(P_s\right) = m_R.\omega^2.y_2\left(y_s\right) \tag{1.13}$$

where:
- m_e, m_w and m_R = Equivalent mass of weightless spindle at point (1) and wharve mass and rotor mass, respectively
- ω = Radian speed of all rotating masses (s)
- y_1 and y_2 (y_s) = Linear displacements of point (1) and the wharve and rotor, respectively

1.3.5 Weightless Shaft

A weightless shaft with a concentrated mass (equivalent mass m_e) will have the following critical speed (natural) frequencies:

a. The linear natural frequency f_n is

$$f_n = \frac{1}{2\pi}\sqrt{\frac{S}{m_e}} \quad \text{s} \tag{1.14}$$

where:
S = Spring constant $(=1/a_{11})$ (see Table 1.1)
m_e = Equivalent mass due to the transfer from a massive shaft to a weightless shaft

b. The critical natural circular frequency ω_n is

$$\omega_n = 2\pi.f_n$$

$$= \sqrt{\frac{S}{m_e}}\ \mathrm{s} \tag{1.15}$$

The critical natural circular frequency ω_n is the critical speed ω_{cr}.

1.3.6 Massive Spindle

The massive spindle has an infinite number of critical speeds $(\omega_{cr_1}, \omega_{cr_2}, ..., \omega_{cr_n})$, while the weightless shaft with a concentrated mass (m_e) has only one critical speed ω_{cr_1}.

1.3.7 Weightless Shaft

The weightless shaft, where it is loaded by a rigid body such as a rotor, will give two critical speeds: the first critical speed ω_{cr_1} is due to centrifugal force and the second critical speed ω_{cr_2} is due to the gyroscopic moment, which is usually too high.

1.4 Active Example

Calculate the critical speeds of a traditional rotor set using the following data:

A. Rotor

$$I_o = 29 * 10^{-6}\ \mathrm{kg/m^2}$$

$$I_q = 20 * 10^{-6}\ \mathrm{kg/m^2}$$

$$m_R = 7.03 * 10^{-2}\ \mathrm{kg}$$

$$K_2 = 1.64 * 10^{-2}\ \mathrm{m}$$

B. Wharve

$$J_o = 2.32 * 10^{-6}\ \mathrm{kg/m^2}$$

$$J_q = 5.8 * 10^{-6} \text{ kg/m}^2$$

$$m_w = 4.8 * 10^{-2} \text{ kg}$$

$$K_1 = 1.36 * 10^{-2} \text{ m}$$

C. Spindle

$$a = 2.22$$

$$\ell_1 = 10^{-2} \text{ m}$$

$$\ell_2 = 4 * 10^{-2} \text{ m}$$

$$E = 206 * 10^9 \text{ pa}$$

$$d = 10^{-2} \text{ m}$$

D. Influence coefficients

$$a_{11} = 1.62 * 10^{-6} \text{ m/N}$$

$$\alpha_{11} = b_{11} = 1.87 * 10^{-6} \text{ N}^{-1}$$

$$\beta_{11} = 2.26 * 10^{-4} \text{ N}^{-1}.\text{m}^{-2}$$

Solution*
Spindle (ω_s)

$$\ell = l_1 + l_2 / 2 \ \left(\text{half spindle length}\right)$$
$$= 10^{-2} + \tfrac{1}{2} * 4 * 10^{-2}$$
$$= 0.03 \text{ m}$$

$$2\ell = 0.06 \text{ m} \ \ (\text{total spindle length})$$

$$I = \frac{\pi d^4}{64}$$
$$= 5.0342 \times 10^{-10} \text{ m}^4$$

$$m_1(m_e) = \rho \times A \times 2\ell (= M) \times 0.236$$

$$= 7800 \times \frac{\pi(10^{-2})^2}{4} \times 0.06 (= M) \times 0.236$$

$$= 0.0368 \, \text{kg} (= M) \times 0.236$$

$$= 8.68 \times (E-3) \text{too small}$$

$$m = \rho \times A$$

$$= 0.613 \, \text{kg/m}$$

$$\therefore \omega_s = \frac{a^2}{\ell^2} \sqrt{\frac{EI}{m}}$$

$$= \frac{(2.22)^2}{(0.03)^2} \times \sqrt{\frac{206 \times 10^9 \times 5.0342 \times 10^{-10}}{0.613}}$$

* The equivalent mass $m_1(m_e)$ can be neglected.

$$\omega_s = 71224.9 \, \text{s}$$

$$\cong 71,225 \, \text{s}$$

Rotor
The constants A, B, C, and D are calculated as follows:

$$A = a_{11}(0 + m_R) + k_2 \alpha_{11} M_R$$

$$= 1.62 \times 10^{-6} (7.03 \times 10^{-2}) + 1.64 \times 10^{-2} \times 1.87 \times 10^{-6}$$

$$= 1.5886 \times 10^{-9} \text{s}^2$$

$$B = \alpha_{11} \times \left(I_o - I\sqrt{q}\right) - a_{11} \, m_R \, k_2 - \alpha_{11}.m_R.K_2^2$$

$$= 1.87 \times 10^{-6} (29.10^{-6} - 20.10^{-6}) - 1.62 \times 10^{-6} \times 7.03 \times 10^{-2}$$

$$-1.87 \times 10^{-6} \times 7.03 \times 10^{-2} \times (1.64 \times 10^{-2})^2$$

$$= -9.90 \times 10^{-11} \text{s}^2/\text{m}$$

$$C = b_{11} * (0 + m_R) + k_2 \beta_{11} + M_R$$

$$= 1.87 \times 10^{-6} \times 7.03 \times 10^{-2} + 1.46 \times 10^{-2} \times 2.26 \times 10^{-4} \times 7.03 \times 10^{-2}$$

$$= 3.920 \times 10^{-7} \text{s}^2/\text{m}$$

$$D = \beta_{11}\left(I_o - I_q\right) - b_{11}m_R k_2 - K_2^2\beta_{11}.m_R$$

$$= 2.26\times10^{-4}\left(29\times10^{-6} - 20\times10^{-6}\right) - 1.87\times10^{-6}\times7.03\times10^{-2}$$

$$\times1.64\times10^{-2} - \left(1.64\times10^{-2}\right)^2 2.26\times10^{-4}\times7.03\times10^{-2}$$

$$= 6.4496\times10^{-9}\,s^{-2}$$

The formula for the critical speed of the rotor is

$$\omega^4\left(B.C - A.D\right) + \omega^2\left(D - A\right) + 1 = 0$$

By substituting the values of the constant, then

$$\therefore \omega_R^4\left(-38.808\times10^{-18}\right) + \left(0.7483\times10^{-18}\right) + \omega_R^2\left(-8.0382\times10^{-9}\right) + 1 = 0$$

$$\therefore -38.808\times10^{-18}\omega_R^4 - 8.0382\times10^{-9}\omega_R^2 + 1 = 0$$

$$\therefore -38.808\times10^{-9}\omega_R^4 - 8.0382\omega_R^2 + 10^{+9} = 0$$

$$\therefore \omega_R = 9352.42\,s$$

$$\doteq 9352\,s$$

Wharve

By using the same technique as used for the rotor, we can write the following constants for the critical speed formula of the wharve:

$$A = 4.4\times10^{-9}\,s^2$$

$$B = -3.330\times10^{-11}\,s^2/m$$

$$C = 5.11\times10^{-7}\,s^2/m$$

The formula for the wharve's critical speed is

$$-8.3\times10^{-18}\omega_w^4 - 6.37\times10^{-9}\omega_w^2 + 1 = 0$$

$$\therefore \omega_w = 11,558\,s$$

To find the resultant critical speed of the traditional rotor set, we apply the Dynkerly formula:

$$\frac{1}{\omega_t^2} = \frac{1}{\omega_s^2} + \frac{1}{\omega_R^2} + \frac{1}{\omega_w^2} \tag{1.16}$$

By substituting to calculate the total speed ω_t, we find $\omega_t = 7240$ s.

$$\therefore n_t = \omega_t \times \frac{60}{2\pi}$$

$$= 69,181 \text{ RPM}$$

(1.17)

The running speed N_R is

$$N_R = 0.7 \times n_t$$

$$= 48,427 \text{ RPM}$$

$$\doteq 50 \text{ kRPM}$$

or

$$N_R = 1.4 \times n_t$$

$$= 96,853 \text{ RPM}$$

$$\doteq 100 \text{ kRPM}$$

The running speed (working speed) of the rotor spinning machine must be 48,427 RPM to avoid the resonance region.

1.5 Summary Points

1. The rotor unit of the rotor spinning machine consists of (a) a spindle (short shaft) with a diameter of φ10 mm, which carries the rotor at one end and the wharve (driving pulley) at the other. (b) Two bearings are located between the rotor and the wharve around the spindle. These two bearings are specially constructed where the inner races are channels on the outer surface of the spindle. The outer races are channels on the inner surface of the gland that embodies both of the antifriction bearings. (c) A gland that surrounds the two single-row radial ball bearings, that is, it is the housing for the rotor unit's bearings.

2. The spindle is an elastic body that can be deflected and returned to its original position. Both the rotor and wharve are rigid bodies that are not deformed during the running of the rotor unit. The laws of solid mechanics can be easily applied to both the rotor and the wharve.

3. The rotor's dynamics (vibrations) refer to the mean amplitude of vibrations, the critical speed, and the dynamic reactions in the rotor's bearings.

4. The dynamic loads on the rotor of the rotor spinning machine are the inertia force (centrifugal force) and the gyroscopic moment.

5. The inertia force is created due to the static unbalance in the rotor's rigid body, and it can be expressed as follows:

$$IF(cF) = M.\omega^2.e \text{ N} \tag{1.18}$$

where:
M = Rotor's mass in kilograms
ω = Radian speed of the rotor $= N \times 2\mu/60$
N = Rotor's RPM
e = Rotor's center of gravity shift from the axis of rotation; in practice, its values are in micrometers (μm), but these must be substituted in meters to have the IF (cF) in newtons (N). The rotor eccentricities are created due to the uneven distribution of the rotor mass around its geometrical axis

6. The gyroscopic moment (M_g) is created due to the deflection of the spindle (elastic body) and the existence of angle φ (see Figure 1.5).

The value of the gyroscopic moment is calculated by the formula:

$$M_g = (I_0 - I_q)\omega^2.\varphi \quad \text{N·m} \tag{1.19}$$

where:
M_g = Gyroscopic moment
I_0 = Polar mass moment of inertia of the rotor's mass
I_q = Equatorial mass moment of inertia around an axis perpendicular to the axis of rotation
φ = Angle between the tangent to the elastic line (deflection curve) of the spindle and the axis of rotation

7. The engineered massive geometric characteristics of the rotor are greater than the wharve; therefore, the reaction in the bearing that is located near the rotor is greater than that in the antifriction bearing of the wharve.

8. The influence (effective) coefficients a_{11}, α_{11}, b_{11}, and β_{11} of the spinning rotor are illustrated in Figure 1.4. They are required for solving

problems of solid mechanics concerning the mechanics of the engineered massive geometrical characteristics of a rotor spinning machine's elements.

Review Questions

Q#1: Describe the constructional aspects of the rotor unit of the rotor spinning machine. Use line diagrams for your explanation.

Q#2: Based on scientific definitions of solid mechanics, select the elements that are rigid bodies in the construction (design) of the rotor unit.

Q#3: With the aid of line diagrams, show the construction of the opening device of a rotor spinning machine.

Q#4: Write a formula for calculating the inertia force on the rotor (cup/pot/camera) of a rotor spinning unit.

Q#5: What is the mathematical relationship between the gyroscopic moment M_g and both the polar and equatorial mass moments of inertia of the rotor's spinning unit?

Answers to Review Questions

Q#1: See Summary Point 1 and Figure 1.1a through c.

Q#2: The rigid bodies are the rotor and the wharve.

Q#3: There is a similarity between the construction of both the rotor unit and the opening device of the rotor spinning machine when the rotor is replaced with a saw-toothed roller.

Q#4: See Equation 1.18 and summary points.

Q#5: See Summary Point 6 and Equation 1.19.

Review Problems

Q#1: Calculate the inertia force on the rotor of a rotor spin box if the rotor's mass $m_R = 73.0$ g, rotational RPM = 70 kRPM, and the radial shift of the rotor's mass center from the axis of rotation = 80 μm.

Q#2: For a rotor spinning machine, what is the value of the gyroscopic moment, M_g, that is created during the rotation of its rotor with a speed = 100 kRPM, a polar mass moment of inertia $I_0 = 3 \; 10^{-6}$ kg/m², and an equatorial mass moment of inertia $I_q = 21 \; 10^{-6}$ kg/m². Assume φ = 0.5°.

Q#3: With the data from Problems 1 and 2, find the total load on the rotor's rigid body.

Answers to Review Problems

Q#1:

$$IF(cF) = M\omega^2.e$$

$$= 70\times10^{-3}\times\left(\frac{70\times2\pi\times10^3}{60}\right)\times(80\times10^{-3})$$

$$= 0.07\times53734513\times0.08$$

$$= 300.913\,N$$

$$= 301\,kN$$

Q#2:

$$M_g = (I_0 - I_q).\omega^2.\varphi$$

$$= (30\times10^{-6} - 21\times10^{-6})\times\left(\frac{100\times10^{-3}\times2\pi}{60}\right)^2\times\left(\frac{0.5\times\pi}{90\times2}\right)$$

$$= 9\times10^{-6}\times109662271\times8.7222\times10^{-3}$$

$$= 8.6084\,N\cdot m$$

$$\cong 9.0\,N\cdot m$$

Q#3: Add the answers for Q#2 and Q#3 to obtain the total load on the rotor of the rotor spinning machine during its operation.

Bibliography

Artobolevski U.U., 1973, *Theory of Mechanisms*, Nauka, Moscow, RFU.

Bevan T., 1969, *The Theory of Machines*, Longmans, London.

Broch J.T., 1980, *Mechanical Vibrations and Shock Measurements*, Larson & Sons, Copenhagen, Denmark.

Elhawary I.A., 1978, Vibrations of heavy twisting spindles, PhD thesis, Moscow State Academy of Textiles (MSAT), Moscow, RFU.

Elhawary I.A., 2012, Vibrations of ring spindles, rotors and drafting systems, Notes, Mansoura University, Mansoura, Egypt.

Elhawary I.A., 2013, Mechanics of the rotor spinning machine, Lecture notes, TED, Alexandria University, Alexandria, Egypt.

Elhawary I.A., 2014, Dynamic balancing of textile rotating masses, Post-graduate course, TED, Alexandria University, Alexandria, Egypt.

Gorman D.J., 1975, *Free Vibrations Analysis of Beams & Shafts*, John Wiley & Sons, Toronto, Canada.

IRD ENTEK, 1996, Introduction to vibrations technology, IRD Mechanalysis, Inc., Columbus, OH.

IRD Mechanalysis Company Inc., 1985, Methods of vibration analysis, Technical Report No. 105, Columbus, OH.

IRD Mechanalysis Inc., 1982, Audio-visual customer training, Special Instruction Manual (Blue Book), Special communications, Alexandria, Egypt.

Karitiski Ya.I., Kornev I.V., Lagynov L.Fy., Cyshkova R.I., and Khodekh M.I., 1974, *Vibrations and Noise Control in the Textile and Light Industries*, Moscow Design Press, Moscow, RFU.

Makarov A.I., Karitiski Ya.I., Andreev O.P., Gladkov K.M., Martirosov A.A., Mulman B.V., and Radyshincky L.A., 1969, Design and construction of the used machines in yarns production, Moscow State Academy of Textiles (MSAT), Moscow, RFU.

Makarov A.I., Karitiski Ya.I., Andreev O.P., Gladkov K.M., Martirosov A.A., Mulman B.V., and Radyshincky L.A., 1976, *Basics of Textile Machines Design*, Machine Design Press, Moscow, RFU.

MIT OPENCOURSEWARE. Mechanical Engineering. http://ocw.mit.edu/courses/mechanical-engineering/, accessed March 3, 2012.

Panovka R.G., 1976, *Bases of Applied Theory of Vibration & Shocks*, MIR Press, Moscow, RFU.

Philipov A.P., 1970, *Vibrations of Elastic System*, Machine Design Press, Moscow, RFU.

Popov A.P., 1975, *The Basic Theory in Design and Construction of the Textile Machines*, Machine Design Press, Moscow, RFU.

Sevetlitcki A.P. and Cmasenko I.V., 1973, *Solved Problems in Vibrations,* Light Industry Press, Moscow, RFU.

Sharma C.S., 1983, *Mechanical Vibrations Analysis*, Khanna Publishers, Delhi, India.

Shiptilnikova A.V., 1975, *Principles of Accurate Engineering, Balancing Part I, Balancing of Rigid Rotors & Mechanisms*, Machine Design Press, Moscow, RFU.

Shiptilnikova A.V., 1975, *Principles of Accurate Engineering, Balancing Part II, Balancing of Elastic Rotors & Balancing Machines*, Machine Design Press, Moscow, RFU.

Yazied T.G., Elyazid T.A., and Doghem M., 2011, Mechanics of Machines (1), Design & Production Engineering Department, Ain Shams University, Cairo, Egypt.

Reference

Makarov A.I., 1968, *Construction & Design of Machines Used in Yarn Production*, Machines Design Press, Moscow, RFU.

2

Rotor Bearings

2.1 Introduction

The spindle (shaft) that carries the rotor and the wharve must be borne on (carried by) a machine's elements, that is, the bearings. Usually, the rotor's bearings are of the short type. The spindle is supported in its location by two ball bearings; one bearing location is near the backside of the rotor, while the other is fixed behind the wharves. Both are single-row radial bearings, that is, the inner race of the bearing is a grooved channel, tracked in the shaft body, while the outer race is another channel grooved and tracked in the inner surface of a gland that surrounds both bearings. The space between the spindle and the gland is filled with a special type of grease as a lubricant, which is paramount to minimizing friction, and it also plays the role of a cooler to a certain extent.

2.2 Ball Bearing

2.2.1 Rigidly Mounted Rotor

Figure 2.1 shows the construction of a commercial rotor set. It includes (1) the rotor body; (2) spindles; (3) outer races; (4) balls; (5) washers; (6) fixer coil springs; and (7) wharves. From Figure 2.1, it is shown that the inner race of the ball bearings of the rotor set is tacked into the spindle surface. Also, the ball bearings' outer races are grooves in the internal surface of the gland where they act as rotor set fixers with the rotor spinning machine's chassis and also store grease lubricant. The residual static unbalance in the rotor's body after manufacturing, plus the unbalance of the deposited trash and fibers on the collecting surface of the rotor, will lead to the creation of an inertia force when the rotor rotates. Under the effect of the centrifugal force, the spindle will be deflected to create angle Ø and consequently a gyroscopic moment is generated. Both the force and the moment, when the rotor rotates, will create dynamic reactions in the bearings. The reaction will generate friction forces where heat is created.

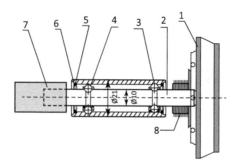

FIGURE 2.1
Rigidly mounted rotor. (1) Rotor body (cup); (2) spindle (inner race); (3) gland (outer race); (4) balls; (5) washer; (6) coil spring; (7) wharve; (8) finned base (cooler).

The created heat can cause the lubricant (grease) to fail and consequently dry friction may occur. To minimize the heat effect, partial cooling takes place by the finned base of the rotor (backside hub) or by the lubricant plus heat dissipation from the rotor set body. Despite these factors, the rotor set with ball bearings cannot run for more than 30 kRPM (production speed). Its lifespan is about 12,000 h, that is, about two years (working hours per

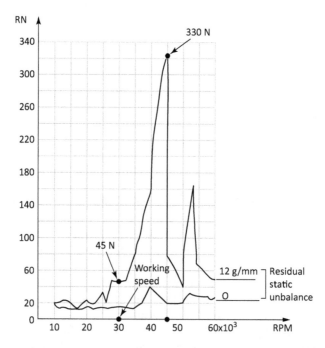

FIGURE 2.2
Rotor's dynamics reaction vs. rotor RPM. (From Moscow State University of Design & Technology (msta.ac), 1975, Rotor's machines, Notes Lab, Moscow, RFU.)

year = 6,000 h). If the dynamic reactions are measured with increasing rotor speed, we obtain the graph in Figure 2.2. The vertical axis indicates the value of the dynamic reaction RN in newtons (N) while the horizontal axis gives the value of the rotor revolutions per minute (RPM). It is clear that the residual static unbalance of 12 g/mm gives a higher value of RN with respect to a 0 g/mm static unbalance. Also, the shape of the graph is too similar to the resonance curve of the rotor. The highest value of RN = 330 N at a rational speed of 45 kRPM—a rigidly mounted rotor's critical speed (Elhawary, 2014).

2.2.2 Elastic Mounting of Rotor

The elastic mounting of the rotor spinning machine's rotor is shown in Figure 2.3. The main gland (socket) (5) is surrounded by another short metallic sleeve (gland) (4) and a long elastic sleeve (3), and these are all embedded by a socket (2) that fixes the rotor set to the spinning machine's chassis. The fitting of the elastic elements in the construction of the rotor set will act as dampers to the stresses in the rotor's bearings due to the reactions; therefore, the value of the reactions is reduced as shown in Figure 2.4. In Figure 2.4, it is clear that when the static unbalance (Me) is reduced from 30 to 0 g/mm, the graphs become smoother. For an elastically mounted traditional rotor, the following mechanical changes have taken place: the working rotor's speed has changed from 30 kRPM (rigidly mounted) to 45 kRPM, and the lifespan of the ball bearings of the rotor's spindle is 30,000 h, or five working years (one working year = 6,000 h). Figure 2.5 shows another type of elastic element that is embedded in the construction of the rotor set. This new type of elastic that wraps around the main gland could improve the smoothness of the dynamic reactions.

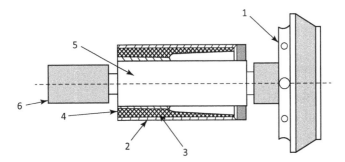

FIGURE 2.3
Elastic mounting of rotor via elastic gland. (1) Rotor body; (2) socket (gland); (3) elastic socket; (4) internal gland (metallic); (5) main gland (outer race of balls); (6) wharve. (From SKF Group, 2013, Roller Bearings Catalogue, Sweden.)

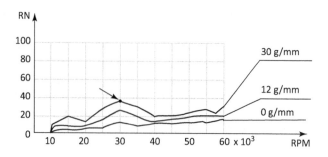

FIGURE 2.4

Rotor bearing dynamic reactions (RN) vs. rotor's speed. M.e: static unbalance massing; e: unbalance mass eccentricity in g/mm, with different values (0, 12 & 30 g/mm); RN: dynamic reaction in bearings (ball) in N; RPM: rotational rotor's speed in revolutions per min.

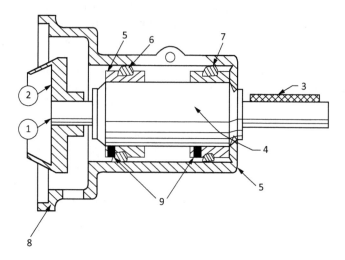

FIGURE 2.5

Rotor's elastic mounting via elastic rings. (1) Rotor spindle; (2) rotor body; (3) tangential belt drive; (4) main gland (outer race); (5) elastic rings holder; (6, 7) elastic rings; (8) rotor housing; (9) screws (fasteners). (From Lawrence C.A. and Chen K.Z., 1984, Rotor spinning, *Textile Progress*, 13/4, 54–58.)

2.3 Magnum Bearing

Figure 2.6 shows the construction of the magnum bearing of a rotor in a rotor spinning machine. The bearing is composed of a small oil reservoir (1), fitted by a porous plug (2) that can be saturated with oil to be dispersed downward as mist oil through the rotor's spindle (3). The misted oil travels right and left to oil the two ball bearings (4) and (5). This lubrication runs continuously as long as the rotor is rotating. The misted oil plays two roles: a lubricant and a cooler, that is, it reduces friction in the ball bearings and reduces

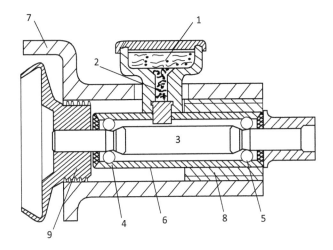

FIGURE 2.6
Magnum bearing. (1) Oil reservoir; (2) porous plug; (3) rotor spindle; (4, 5) ball bearings; (6) main gland–socket–outer race; (7) rotor housing; (8) elastic sleeves; (9) rotor finned base. (From Lawrence C.A. and Chen K.Z., 1984, Rotor spinning, *Textile Progress*, 13/4, 54–58.)

temperature increases. The mechanism of oil misting and dispersing from upward to downward can be explained as follows: the rotation of the rotor's spindle forces the air in the gap between the spindle and the gland (outer race) (8) to be displaced outward via the bearings; a vacuum will be created as the escaped air is substituted by the mist oil from the tank (1) through the porous plug (2). The incorporation of elastic sleeves (8) in the rotor set corpus improves the rotor dynamics via a reduction in reactions and vibration noise. Consequently, the following mechanical relative merits are obtained: (1) an increase in the working speed of up to 60 kRPM; (2) the assurance of continuous lubrication for ball bearings (avoiding dry friction); (3) continuous active cooling because of the instantaneous fresh oil mists; and (4) low levels of vibration and noise, thus reducing costs. The application of grease as a lubricant is not preferable because of its transition from a solid state to a liquid state and consequently its evaporation due to the generation of heat in the bearings; this evaporated liquid grease will pollute the environment and will create dry friction in the ball bearings of the rotor. In general, the new mist oil technology is highly appreciated (Elhawary, 2013).

2.4 Cooling of the Rotor's Ball Bearing

At running speed, the rotor will be loaded by inertia force and gyroscopic moment that will create dynamic reactions in the rotor's bearing. The bearing's reactions will be transferred to frictional forces and, consequently, heat generation in the whole rotor set will be established. The generated heat will change the boundary lubrication to dry friction, which will be forcibly burned and expelled from the grease, meaning the bearing's failure is assured. Therefore, a cooling action must be devised and applied by wrapping metallic rings around the main gland of the rotor set (Figure 2.7), manifesting the rotor with a finned base (hub) to increase its surface area subject to room temperature, minimizing the load on the rotor via dynamic balancing to minimize bearing friction, and finally enabling the effective tracking of the air flow via the fiber transport tube to inside the rotor and then into the spinning room's atmosphere.

2.5 Rotor Air Bearing

The concept of the air bearing depends on the clearance between the rotor's spindle and the main gland that surrounds the spindle's bearings. In the clearance, compressed air is pumped under a pressure of 4000–8000 kPa, then the spindle of the rotor will float, that is, there will be no metallic contact and consequently no bearing friction. The problem in such a bearing

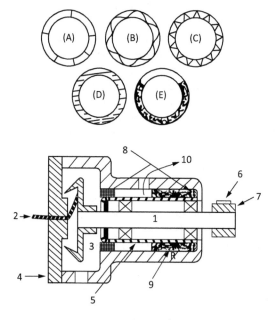

FIGURE 2.7

Rotor bearings cooling. A–E: Metallic elements with good heat conduction embedded in the rotor set design; (1) rotor spindle; (2) rotor's spun yarn; (3) rotor body; (4) rotor housing; (5) bearing gland; (6) belt; (7) wharves; (8) incorporated metallic elements; (9) elastic sleeves; (10) expelled hot air.

is at the start or end of the rotor where there is metallic contact. Also, the air consumption is too great and the air must be clean. In addition, if the value of x clearance is too small at 5–17 mm, the rotor could fail at any time. But some advantages can also be obtained; for example, the running speed of the rotor is unlimited when there is no bearing friction. The only limitation is the rotor's bursting strength. The air bearing is considered a user-friendly item.

2.6 Twin Disc Bearings

This type of rotor bearing (Figure 2.8) has been introduced to the spinning industry via German machine makers Sussen. The rotor can run with a speed of more than 100 kRPM. The design of the bearing has two similar adjacent discs whose outer circumferences are covered by elastic frictional material made from polyurethane that is prepared by chemical cooking. It is named Vulkollan covering or polyurethane covering. The frictional properties of the covering help the stability of the rotor. As the rotor needs two supporting points (surfaces), another two discs must be placed behind the two previously mentioned. The elasticity of the cover plays a role in vibration damping from the noise and consequently the dynamic reactions can be reduced. The Twin Disc bearing construction means that it has four discs similar to each other, two on the front side and two at the back side; all discs are running with the same RPN and consequently the same surface speeds (all discs have the same diameters). Also, it means that there are two shafts,

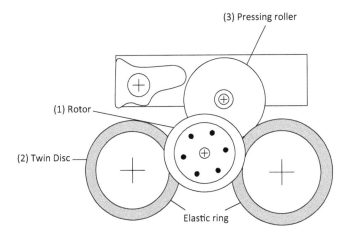

FIGURE 2.8
General front view of a Twin Disc bearing. (1) Rotor corpus; (2) front view of two discs; (3) pressing roller of rotor spindle's driving belt. (Modified from Spindelfabrik Sussen, Schurr, Stahlecker & Grill GmbH, Sussen, Wurttemberg, Germany. Private communication.)

one for each pair of discs. These shafts are not parallel, but their axes could be intersected to form the letter V (wide at back discs and narrow at front discs) so that the rotor's spindle is not horizontal but slightly inclined to the back side. It will then be subjected to a thrust reaction during running. The Twin Disc A bearing needs a thrust bearing to absorb the thrust reaction. Figure 2.8 shows a general front view of the Twin Disc bearing, while an isometric drawing is shown in Figure 2.9. The relative merits of the Twin Disc bearing are that the rotor must be changed easily according to the processed fiber's staple length where the rotor's diameter is connected technologically by the fiber staple length according to the formula:

$$D_R \cong 1.4 \times L_s \qquad (2.1)$$

This means that the rotor must be changed when the processed fibers are changed. This is achieved easily via Twin Disc bearing version rotors with different diameters, which are illustrated in Figure 2.10. Thus, the rotor must be changed according to the fiber length.

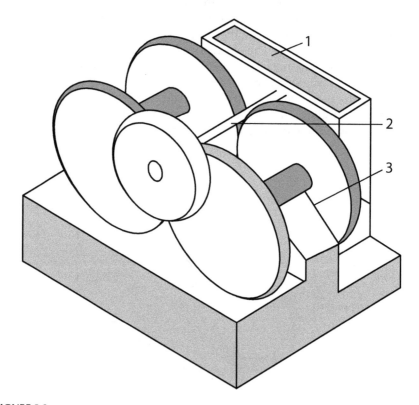

FIGURE 2.9
Isometric drawing of a Twin Disc bearing. (1) Disc covered by elastic material (polyurethane or Vulkollan covering); (2) rotor spindle; (3) bracket for Twin Disc shaft.

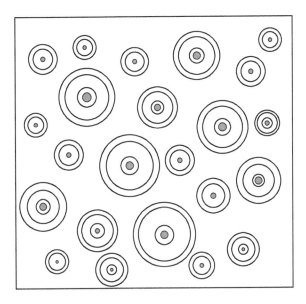

FIGURE 2.10
Versions of rotors with different diameters.

Figure 2.11 shows the relationship between the power requirements in watt per rotor and the rotor's speed in kRPM. For different rotor diameters, it shows that the increase in the rotor speed from 10 to 110 kRPM will increase the power requirement per rotor from 20 to 200 W per rotor. Also, the increase in the rotor diameter 0 from 40 to 86 mm will increase the power requirement per rotor. The most popular rotor diameters (φ) in rotor spring machines are 56, 46, and 40 mm. Nowadays, we have $\varphi = 33$ mm and sometimes 28 mm. For the three popular diameters, the average power requirement per rotor is $\cong 70$ W. For such power as shown in Figure 2.11, the running speed of the rotor will be 50, 60, and 70 kRPM, respectively. With such increases in rotor speeds, the energy saving for 1 kg of rotor-spun yarn will be 13% for rotor $\varphi 46$ mm and 21% for rotor $\varphi 40$ mm; we can conclude from all the previous figures concerning the rotor change due to fiber length that an energy saving from 13% to 21% for the production of 1 kg of rotor-spun yarn can be achieved. This is a good relative merit of the Twin Disc bearing.

Figure 2.12 shows the required centrifugal (inertia) force required to collect the fibers on the rotor collecting surface to form the yarn (metric yarn count is 34). It is clear that the centrifugal force in centinewtons (cN) increases with the increase of the rotor speed in kRPM. Practically, it is found that the value of such an inertia force is 38 cN for yarn (metric count 34) and to reach such force we need to apply a rotor speed of 50 kRPM for a rotor diameter of 56 mm, while we also need to rotate the rotor by 70 kRPM for rotor $\varphi = 40$ mm. This will also help in energy saving during rotor-spun yarn production, and is a further advantage of the Twin Disc bearing. Another relative

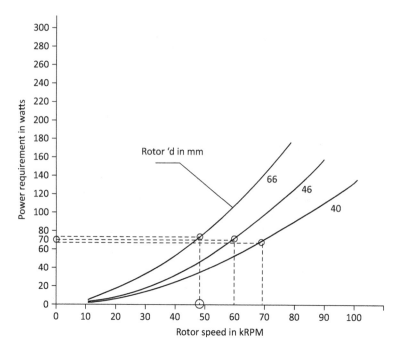

FIGURE 2.11
Power requirements for rotor speed.

merit of the Twin Disc bearing is the low noise level because no balls roll at relatively high speeds. In addition, the discs' circumference is covered with elastic material that absorbs the rotor's vibration and noise. Lastly, the Twin Disc rotates at a relatively low speed with respect to the rotor itself. Beside these items, the Sussen factory embodies every 12 rotors with their Twin Disc bearings in a closed channel, the inner wall of which is covered by absorbent noise material. The Twin Disc bearing is considered a user-friendly item with respect to noise and, to a certain extent, with respect to air pollution (lubricant vapor).

2.7 Thrust Bearing

For rotor stability in the Twin Disc bearing, it was necessary to make the Twin Discs' shafts not parallel, so they were intersected as shown in Figure 2.13. Therefore, the rotor's spindle needs a thrust bearing at the rear end of the rotor's spindle to absorb the axial reaction from the rotor body itself and from the driving tangential belt. There are many designs for the thrust bearings of the rotor's spindle. Figure 2.14a shows that at the near end of the rotor's spindle (1), there is a fractional conical tapered end (2) that contacts continuously with another frictional conical roller (5), which prevents the

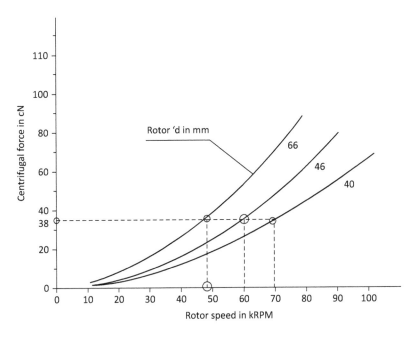

FIGURE 2.12
Centrifugal force in cN required to produce yarn of metric count 34 with rotor speed in kRPM for various rotor diameters φ86–40 mm.

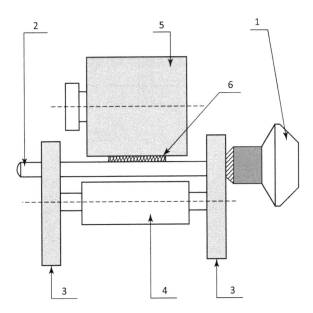

FIGURE 2.13
The Twin Discs' axes are not parallel. (1) Rotor corpus; (2) rotor spindle; (3) Twin Discs; (4) Twin Discs bearings socket; (5) tangential belt press roller; (6) tangential belt.

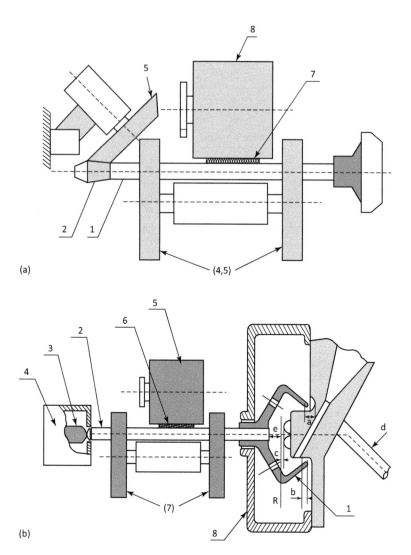

FIGURE 2.14

(a) Frictional conical roller as thrust bearing for the rotor spindle. (1) Rotor spindle; (2) rotor spindle tapered frictional end; (1,2) front two discs; (3,4) back two discs; (1,2) and (3,4) Twin Disc bearing; (5) frictional conical tapered roller; (6) belt drive; (7) belt drive pressing roller. (b) Industrial rotor set with a free rotating ball in an oil bath. (1) Rotor (Y-type); (2) rotor spindle; (3) free rotating ball (thrust bearing); (4) an oil bath—trough; (5) belt drive pressing roller; (6) belt drive; (7) Twin Disc bearing; (8) rotor housing.

axial movement of the rotor's spindle. An industrial rotor set with a thrust bearing as a ball rotates in an oil bath (trough) that acts as a continuous lubricant for the ball to minimize wear action on the ball, as in Figure 2.14b. Also, the technological settings of the rotor set have been illustrated to have optimal rotor-spun yarn quality (Elhawary, 2012).

Rotor's technological settings

1. Setting of rotor's edge with regard to housing edge
2. Setting of fibered tube with regard to rotor edge
3. Setting of yarn tube with regard to rotor collecting surface
4. Yarn tube's navel (hub)
5. Setting of navel with regard to rotor base
6. Rotor's collecting surface

2.8 Summary Points

1. Seemingly, the most applicable antifriction bearing in the rotor spinning boxes are single-row radial ball bearings.
2. For rotors with traditional antifriction bearings, that is, rigidly mounted rotors, the working speed is 20 kRPM.
3. The incorporation of elastic elements in the construction of the rotor's unit such as elastic glands or elastic rings around the bearing glands (bearing housing) increases the rotor running speeds to 60 kRPM, especially when using misting oiling systems (magnum bearing).
4. The loading on the rotor will generate inertia force and a gyroscopic moment in the rotor's spindle bearings during the spinning process.
5. The bearings of the rotor spindle are the carriers for the dynamic reactions while the rotor is working.
6. The bearings that are located near the rotor (cup, pot, and camera) are carrying more dynamic reactions than the bearings that are located near the wharve. The massive geometrical characteristics of the rotor (rigid body) are relatively greater than the wharve.
7. The compound of massive geometrical characteristics of movable parts in rotor spinning needs the influence coefficients, as helping tools for studying their dynamics.
8. The Sussen Twin Disc bearing is highly respected due to its numerous relative merits as an energy saver and its ability to lower the noise level of the spun box of the rotor spinning machine.
9. The centrifugal force between the rotor spindle bearings' dynamic reaction and the rotor RPM is too similar to the resonance curves of the same rotor.
10. The rotor diameter has a big effect on the rotor power requirements. As the rotor diameter increases, the power requirement increases significantly. As a rough figure, the power requirement of the rotor

irrespective of its diameter and taking practical considerations into account is 70 W (ω).

11. The breakdown of the lubricant film inside the bearing will lead to metallic contact and consequently heat generation. With poor cooling of the rotor box, the energy spike will increase rapidly and bearing failure will occur.

12. The cooling of the spinning unit is obtained via the finned base of the rotor, the metallic rings around the rotor spindle's bearings gland, more openings for ventilation, air flow inside the unit during fibers withdrawal in their duct, and so on.

13. The dynamic reaction in the rotor spindle bearings is about 45 and 20 cN for rigidly mounted and elastically mounted rotors, respectively (static unbalance = 12 g/mm).

14. Some modern spinning rotor machines have filters with a ceramic bearing for high speed.

Review Questions

Q#1: Explain with the aid of line diagrams each of the following: (a) rigidly mounted rotor and (b) elastically mounted rotor.

Q#2: Graph the relationship between the rotor spindle's dynamic reactions and the rotor spindle speed to rigidly and elastically mounted rotors.

Q#3: Explain in detail the Sussen Twin Disc bearing for the rotor spin position. Mention its relative merits.

Q#4: How was the rotor's thrust bearing resisted (beared)? State different concepts.

Answers to Review Questions

Q#1: See Figures 2.1 and 2.3.

Q#2: See Figures 2.2 and 2.4.

Q#3: See Section 2.6 and Figures 2.8 through 2.14b.

Q#4: See Figures 2.13 and 2.14b.

2.9 Related Information to Chapter 2

Nominal bearing life

In choosing a bearing for a rotor spinning machine, the nominal life is one of the essential measures in bearing calculations. The nominal life L is the number of revolution hours at a constant speed attained or exceeded by 90%

of a sufficiently large batch of apparently identical bearings before the first evidence of material fatigue occurs (SKF Group, 2013).

The following list shows the approximate normal life values for some textile machines.

- False twist roll bearings 10 kh
- Rotor spindle bearings 20 kh
- Drafting rolls of drafting systems 75 kh

Dynamic reactions
The calculations of the dynamic reactions' magnitude and effective geometric points are used. The number of revolutions must be recorded. Also, a main value must be determined for the speed if it is not constant.

Dynamic reactions determination:
For each row of rolling elements in the antifriction bearings, the movable and stationary dynamic reactions are combined into an average (mean) value according to the formula:

$$R_m = f_m (R_1 + R_2) \qquad (2.2)$$

where:
R_m = Mean dynamic reaction
R_1 = Static reaction
R_2 = Dynamic reaction
f_m = Coefficient (SKF tables)

2.10 Equivalent Dynamic Load Reaction of the Antifriction Bearing

The radial and the axial dynamic reactions cannot be determined due to the formula:

$$R_e = XR_r + XR_a \qquad (2.3)$$

where:
R_e = Equivalent dynamic reaction (load)
R_a = Actual reaction (axial force)
R_r = Dynamic reaction (radial load)
The foregoing equations and calculating factors are widely used in deep groove bearings.

Determining the normal life:

With the aid of equivalent dynamic reaction R_e and the dynamic load rating (basic) C of a row of rolling elements, its normal life L_{10h} in hours can be determined according to the equation:

$$L_{10h} = \frac{(E+6)}{60.n}\left(\frac{C}{R_e}\right).p \tag{2.4}$$

where:

L_{10h} = Normal antifriction bearing life in hours
n = Journal speed, RPM
P = Exponent, for one row ball bearing $p = 3$
$p = 3.33$
C = Dynamic load rating
h = Hours

It is well known that the probability is higher for bearing failure when it has two rows of rolling elements—if the bearing life is L_1 and L_{11} for single row #1 and single row #2, respectively, then the total nominal life of a bearing with two rows is L, where $1/L = 1/L_1 + 1/L_{11}$.

Here, the value of L by calculation has an 8% deviation as the normal life of a bearing unit is practically always less than the individual lives of the rows of its rolling elements (inside the bearing unit).

2.11 Equivalent Static Load Rating

The static reactions comprising radial and axial components must be converted into an equivalent static bearing load.

The equivalent static load rating is a combination of the actual radial reaction and the actual axial reaction as shown in Equation 2.5. The following equation is applicable:

$$R_0 = X_0.R_r + Y_0.R_a \tag{2.5}$$

where:

R_0 = Equivalent static bearing force that envelopes both R_r and R_a
R_r = Actual radial reaction
R_a = Actual axial reaction
X_0, Y_0 = Radial and axial factors

All the required information for the determination of the equivalent static loads is given in special tables; if the equivalent static load R_0 is less than the radial static reaction R_r, use $R_0 = R_r$.

Requirements of static load rating:
The basic static load rating requirements C_0 of a bearing are determined by the equation:

$$C_0 = S_0 . R_0 \tag{2.6}$$

where:
  C_0 = Basic static load rating
  R_0 = Equivalent static bearing force
  S_0 = Static safety factor

Bearing material hardness is reduced in the high temperature range. Any resultant influence on the static load capacity is mentioned in detail with regard to specific issues.
In general, the following figures are the minimum values that can be applied:

- Smooth vibration: free working $S_0 = 0.5$
- Quiet working for medium running conditions with normal demands $S_0 = 1$
- Shock loads (pronounced) $S_0 = 1.5-z$
- High requirements (demands) on quiet working $S_0 = 2$

For very slow rotating bearings and where the lifespan is short, in terms of number of revolutions, the basic static load rating must be considered as the life equation can be incorrect in giving apparently allowable loads that far exceed the basic static load rating.

Bearing rotation:
For heavy loads variations, the static load rating must consider these variations of shock loads as sudden or impact loads. Besides, accurate calculation of shock load is not guaranteed. The unavoidable good load distribution will make deformation in bearing housing essential. In addition, if shock loading is still effective during several revolutions of the bearing, the even deformation will be noticeable in the race ways and damaged indentations will be avoided.

Bibliography

Artobolevski O.U., 1973, *Theory of Mechanisms*, Nauka, Moscow, RFU.
Bevan T., 1969, *The Theory of Machines*, Longmans, London, UK.
Broch J.T., 1980, *Mechanical Vibrations and Shock Measurements*, Brüel and Kjær, Naerum, Denmark.

Doghemy H. and Taher Y.A., 2011, Mechanics of machines (1), Design & Production Engineering Department, Ain Shams University, Cairo, Egypt.

Elhawary I.A., 1978, Vibrations of heavy twisting spindles, PhD thesis, Moscow State Academy of Textiles, Moscow, RFU.

Gorman D.J., 1975, *Free Vibrations Analysis of Beams and Shafts*, Wiley, Toronto, Canada.

IRD ENTEK, 1996, Introduction to vibrations technology, IRD Mechanalysis Inc., Columbus, OH.

IRD Mechanalysis Company Inc., 1982, Audio visual customer training, Special instruction manual (Blue Book), Special communication, Alexandria, Egypt.

IRD Mechanalysis Company Inc., 1985, Methods of vibration analysis, Technical Report No. 105, Columbus, OH.

Karitiski Ya.I., Kornev I.V., Lagynov L.Fy., Cyshkova R.I., and Khodekh M.I., 1974, *Vibration and Noise in Textile and Light Industry Machines*, Moscow Design Press, Moscow, RFU.

Makarov A.I., Karitiski Ya.I., Andreev O.P., Gladkov K.M., Martirosov A.A., Mulman B.V., and Radyshincky L.A., 1969, Design and construction of the used machines in yarns production, Moscow State Academy of Textiles (MSAT), Moscow, RFU.

Makarov A.I., Karitiski Ya.I., Andreev O.P., Gladkov K.M., Martirosov A.A., Mulman B.V., and Radyshincky L.A., 1976, *Basics of Textile Machines Design*, Machine Design Press, Moscow, RFU.

Panovka R.G., 1976, *Bases of Applied Theory of Vibration and Shocks*, MIR Press, Moscow, RFU.

Philipov A.P., 1970, *Vibrations of Elastic Systems*, Machine Design Press, Moscow, RFU.

Popov A.P., 1975, *The Basic Theory in Design and Construction of the Textile Machines*, Machine Design Press, Moscow, RFU.

Rengasamy R.S., 2002, *Mechanics of Spinning Machines*, NECUTE, Delhi, India.

Sharma C.S., 1983, *Mechanical Vibrations Analysis*, Khanna Publishers, Delhi, India.

Shiptilnikova A.V., 1975, *Principles of Accurate Erg. Balancing, Part I, Balancing of Rigid Rotors and Mechanisms*, Moscow Design Press, Moscow, RFU.

Shiptilnikova A.V. (ed.) 1975, *Principles of Accurate Erg. Balancing, Part II, Balancing of Elastic Rotors & Balancing Machines*, Moscow Design Press, Moscow, RFU.

Svetlickiv A. and Cmasenko I.V., 1973, *Solved Problems in Vibrations*, Vishaya Ahkola, Moscow, RFU.

References

Elhawary I.A., 2012, Vibrations of ring spindles, rotors and drafting system, Notes, Mansoura University, Mansoura, Egypt.

Elhawary I.A., 2013, Mechanics of the rotor spinning machine, Lecture notes, TED, Alexandria University, Alexandria, Egypt.

Elhawary I.A., 2014, Dynamic balancing of textile rotating masses, Post-graduate course, TED, Alexandria University, Alexandria, Egypt.

Lawrence, C.A. and Chen K.Z., 1984, Rotor spinning, *Textile Progress*, 13/4, 54–58.

Moscow State University of Design & Technology (msta.ac), 1975, Rotor's machines, Notes Lab, Moscow, RFU.

SKF Group, 2013, Roller Bearings Catalogue, Sweden.

Spindelfabrik Sussen, Schurr, Stahlecker & Grill GmbH, Sussen, Wurttemberg, Germany. Private communication.

3

Rotor Drives

3.1 Introduction

When the speed of commercial rotor spinning machines increases, the operators of spinning mills will greatly welcome such progress. Rotor spinning machine makers have attained advanced development in new designs for rotor spindle bearings with low noise levels and low mechanical vibrations. Consequently, the lifespan of the bearing has been prolonged. Based on general results, the new Twin Disc type of bearing can run at high speeds.

In general, the rotor spinning machines in the spinning mills are operated via a central driver by one tangential belt (group drive) that occupies a full single-side length of the machine. Here, it must be mentioned that the rotor spinning machine's rotor and transporting elements are installed in one housing at one end of the machine.

When efforts were first made to increase rotor spinning machine speed, the designers of the rotor drive considered applying a single rotor drive, that is, a high-frequency motor for each rotor. The motor's axis is the rotor spindle.

3.2 Individual Rotor Drive (Patented)

As mentioned previously, the individual rotor drive means a single high-frequency motor for each rotor. The rotor spindle is the motor axis. The relative merits of the drive are as follows:

1. Low noise levels due to a low number of antifriction bearings plus less masses of moving air.

2. Low power requirements for each rotor due to the absence of power loss in the tangential belt drive, the absence of tangential belt pressure on the rotor's wharve, the decrease in air conditioning costs, and no loss in power factor ($\cos \varphi$) in the dc motors.

3. The simplicity of the spinning position's construction and the consequent ease with which the spinning unit can be assembled, which has reduced sudden accidents. A high level of flexibility of the individual drive has also been attained.

4. The running cost is reduced due to low maintenance costs and accordingly an increase in the spinning box's productivity.

5. The smoothness of the rotor as it picks up speed (accelerating period).

6. The increase in the rotor spinning machine's efficiency due to the decrease in total breakdown time (low maintenance and repair).

7. The spinning of the rotor-spun yarn makes individual quality control possible.

The drawback of the individual rotor drive is that, until now, it was not available commercially, except for the BD 200 BDA machine. This may be due to

1. The cooling of the spinning unit in the group

   The central drive of all the motors with their associated elements are housed at one end of the machine, so the heat-generating elements are isolated from the rotors.

   In the individual drive, the individual high-frequency motor is a heat generator; therefore, the rotor's operation is greatly affected and consequently the spinning technology. In such drives, an additional cooling action must be used such as the separation of the feed air flow transporting opened fibers from the opening device to the rotor is redirected to the spin box for cooling. In addition, a cooling unit can be incorporated in the individual spinning position to maintain the inside temperature at 20°C, which will provide the best spinning technology.

2. The accuracy of the rotational speed

   If the spinning speed is higher than 30 kRPM, then the individual drive must be far from its resonance, which depends on the constructional aspects of the drive. In general, the maximum spinning speed depends on the yarn megapascal (MPa) and the maximum allowable temperature or both.

   To attain an input-stable nominal frequency, we must use a continuous frequency stabilizer.

3. Energy consumption depends mainly on low noise levels and low energy consumption; air friction or drag, that is, air noise and friction in the antifriction bearings (rotor spindle bearings) are due to the many rolling elements inside the bearings.

A. *Air drag: Friction*

   Irrespective of the slightly rough surface of aluminum (motor's rotator), the motor stator's individual magnetized elements must have smooth surfaces; therefore, these elements with their hysteresis rings must be sprayed and covered with plastic. For the same reason, the rotor stator must not have any open or broken coils. The coils of the stator must also be sprayed with plastic to ensure smooth surfaces. The energy savings due to smoothing the surfaces is 10 W per rotor, and the noise level is also reduced.

When a disc rotates in an open-air space, the air drag will be large. To minimize the air space, a cover can be used for the spinning position. The energy saving from this method is 13 W for a rotor with a diameter of φ48 mm and a speed of 70 kRPM. In addition, the continued reduction in air space around the rotor so that it is less than 0.2–0.4 mm will create difficulties in manufacturing, and energy savings will not offset the manufacturing cost.

B. Bearings friction

The friction from the rotor spindle's antifriction bearings depends on many factors such as spindle diameter (sliding velocity), lubrication gap, viscosity of lubricant, and working temperature.

The rotor spindle diameter has an appreciable effect on bearing friction. A small spindle diameter reduces friction due to the reduction in the sliding velocity, but a diameter that is too small makes manufacturing and handling difficult; therefore, we reduce the diameter as much as is possible yet practical. For an allowable spindle diameter, a lubricant of low viscosity is applied.

To minimize frictional noise, we must decrease the effect of mechanical vibration by incorporating elastic elements in the bearings' housing, in addition to decreasing the mass of the bearing. Elastic elements decrease the stresses placed on the bearing, especially at the start of operation when the rotor passes the resonance region, and resists the transmission of noise from the bearings to the machine chassis.

Figure 3.1 shows the construction of an individual rotor drive for a rotor of a rotor spinning machine. The construction of the drive consists of carrying rollers (4), (5), and (6) (not shown) that form a conical shape, the axis of which coincides with the rotor axis. The surfaces of these rollers form a tetrahedron shape in the space to support the short rotor spindle (conical shaft). The rotor shaft is driven by a roller (4). The surfaces of these rollers are covered by elastic materials. To ensure the smooth running of the rotor, the rotor conical spindle (8) is kept in contact with rollers (4), (5), and (6). At its right or lower end while the other end carries the rotor, the motor shaft (10) of the motor (11) is surrounded by housing (9).

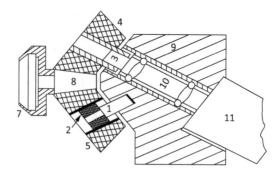

FIGURE 3.1
Individual rotor's drive. (1) Bearing shaft of roller; (2) antifriction bearing of shaft; (3) shaft of roller; (4) supporting roller; (5) supporting roller; (6) unshown supporting roller; (7) rotor: cup, pot, camera; (8) rotor: short conical spindle; (9) motor's housing; (10) motor's shaft; (11) motor. (From Lawrence C.A. and Chen K.Z., 1984, Rotor spinning, *Textile Progress*, 13/4, 54–58.)

A commercial rotor spinning machine BD 200 BDA runs with an individual rotor drive where the rotor can work up to 100 kRPM and the rotor's power requirements are 150–200 W for running speeds 60–90 kRPM, that is, 175 W/rotor.

3.3 Central Group Drive

The central group drive of the rotor spinning machine's rotor consists of a tangential belt. Therefore, such a drive is considered cost-effective. But when it is required to run at high speeds, a lot of problems can occur, such as mechanical vibrations, heat increases, noise level increases, and wear. Most of these problems can be solved through the improvement of the rotor spindle bearings design. Figure 3.2 shows a tangential belt that works via a central group drive. This type of drive is applied commercially in rotor spinning machines. This drive system consists of

1. Pressing or tension rollers that press the belt against the rotor's wharves to increase the contact angle and consequently the friction
2. Rotor wharves that are fixed at one of the rotor spindle's ends

The initial tension of the belt is activated by a coil spring at the drive pulley.

1. The drive pulley works under the effect of a tension spring to create the initial tension to secure continuous belt tension.
2. The drive pulley is driven though the rotor's motor (main motor). Its radian speed ω is in seconds (s).

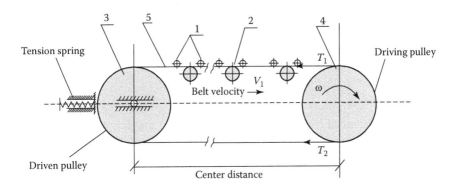

FIGURE 3.2
Rotor's tangential belt drive. Pressing rollers (tension roller) (1); wharves of rotors (2); driven pulley of the tangential belt (3); driving pulley of the tangential belt (4); tangential belt (5); V_1 belt velocity; T_1 = belt tension on tight side; T_2 = belt tension on slack side; ω = radian speed of driving pulley. (From Moscow State University of Design & Technology, Moscow, RFU, 1978.)

3. The tangential belt drives all the rotors for a single-sided machine; the calculation of the belt can be written as follows:

The tension in the tight side T_1 is calculated by the formula:

$$T_1 = T_2 + Z_1 P_1 + Z_2 P_2 + P_3 \tag{3.1}$$

where:
T_1 and T_2 = Belt tension in the tight and slack sides, respectively
P_1 = Belt tension around the rotor's wharve
P_2 = Belt tension around the tension roller
P_3 = Belt tension around the belt's driven pulley
Z_1, Z_2 = Number of rotor and tension rollers, respectively

The tensions P_1, P_2, and P_3 are calculated by the following formula:

$$\left. \begin{aligned} P_1 &= \frac{102\, N_1}{V} \\[6pt] P_2 &= \frac{102\, N_2}{V} \\[6pt] P_3 &= \frac{102\, N_3}{V} \end{aligned} \right\} \tag{3.2}$$

where N_1, N_2, and N_3 are the power requirements to run the rotor, tension roller, and driven pulley, respectively.

Another well-known formula of the ratio between T_1 and T_2 is

$$T_1 = T_2 \times e^{\mu\theta} \tag{3.3}$$

where:

    μ = Coefficient of friction between the belt and the driving pulley
    θ = Central angle of contact between the belt and the driving pulley;
        when both the driving and driven pulley are of equal diameters, the
        value of angle $\theta = \pi$
    e = Constant figure = 2.71

From Equations 3.1 through 3.3, we can obtain

$$\left.\begin{array}{l} T_2 = \dfrac{Z_1 P_1 + Z_2 P_2 + P_3}{e^{\mu\theta} - 1} \\[3ex] T_1 = \dfrac{Z_1 P_1 + Z_2 P_2 + P_3}{e^{\mu\theta} - 1} \times e^{\mu\theta} \end{array}\right\} \tag{3.4}$$

The total stress as a maximum is found from the formula:

$$\sigma_{max} = \sigma_1 + \sigma_v + \sigma_b \tag{3.5}$$

where:

    σ_{max} = Total maximum stress in the belt
    σ_1 = Stress due to tension S_1
    σ_v = Stress due to centrifugal force on the belt when it is passing around
        the pulley circumference.
    σ_b = Stress due to the belt bending around the pulley

The value of these stresses is calculated from the formula:

$$\left.\begin{array}{l} \sigma_1 = \dfrac{S_1}{t \times b} \\[3ex] \sigma_v = \dfrac{\gamma}{g}.v^2 \\[3ex] \sigma_b = \dfrac{t}{d}.E \end{array}\right\} \tag{3.6}$$

where:

    t = Belt thickness
    b = Belt width
    γ = Specific weight of belt material
    g = Gravitational acceleration
    d = Driving pulley diameter
    E = Young's modulus

$$\therefore \sigma_{max} = \frac{\left(z_1 p_1 + z_2 b_2 + P_3\right) \times e^{\mu\theta}}{\left(e^{\mu\theta} - 1\right) \times t \times b} + \frac{\gamma}{g}.v^2 + \frac{t}{d} \times E \leq [\sigma] \tag{3.7}$$

where $[\sigma]$ is the design stress of the belt.

The value of $[\sigma]$ is $\leq$10–15 MPa. The belt width b is calculated by the formula:

$$b \geq \frac{\left(z_1 p_1 + z_2 b_2 + P_3\right) \times e^{\mu\theta}}{t \times (e^{\mu\theta} - 1) \times [\sigma] - \frac{\gamma}{g}.v^2 - \frac{t}{d} \times E} \tag{3.8}$$

The disadvantages of the central group drive are as follows:

- The belt's lifespan is not long.
- High noise level.
- The slip percentage between the belt and the pulleys is 4%–5%.
- In the event of sudden belt failure (breakage), the full rotor is stopped so that the whole machine doesn't work.
- The maximum belt speed $v = 30$–35 m/s, so the rotor's RPM = 32,000–40,000 for a wharve $Q = 18$ mm.

3.4 Indirect Rotor Drive

The indirect rotor drive is very similar to the central group drive except that the traditional single-row radial ball bearings of the rotor spindle are replaced by Twin Disc bearings as carriers. These types of carriers must be applied to the rotor's spinning box (position) when the rotor's speed is higher than 60 kRPM. Consequently, it is well known that indirect rotor drives are suitable for high speed rotors.

The Sussen factory in Germany produced and introduced the multiplying drive and the reducing drive for spinning industry machines.

A. *The multiplying drive of the first-generation rotor*

Figure 3.3 shows the design (construction) of the multiplying rotor drive. It consists of

1. Two front carriers (two front discs). The circumferential surfaces are covered by an elastic material made from polyurethane; the elastic material has frictional properties that help rotor spindle rotation due to continuous contact between them. In addition, polyurethane as an elastic material absorbs (dampens) both mechanical vibrations and noise.

   Also, there are another two discs very similar to the previous ones that are located on the back side of the drive with a shaft that is proportional to the rotor spindle's length.

   The two near wharve rollers are covered by the two front discs (all four discs have the same diameter φ72 mm). This means that the drive has four discs, every two of which are fixed on a small shaft φ9.0 mm in length with a diameter of φ9 mm. At the front end of the discs' shaft, there are two small wharves W that are 20 m in diameter, which drive the two discs via a tangential belt drive that has a linear speed $V_R = 10$ m/s (the same linear speed of small wharves of a Twin Disc spindle).

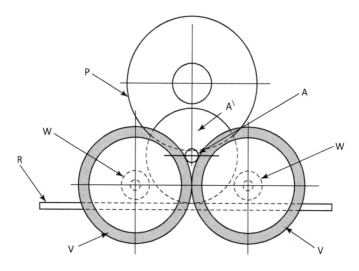

FIGURE 3.3

Indirect rotor's drive—Multiplying drive. V, Vulkollan cover of Twin Disc drive (φ72 mm, 12.5 kRPM); R, tangential belt; W, Twin Disc roller wharves (φ20 m); A, cross-sectional area of rotor spindle φ = 9.0 mm; A', rotor (cup, pot, camera) (100 kRPM); P, pressing roller (tension roller). (From Sussen Group, ITMA, 1971, Rotor speeds up to 80.000 rpm, Special Catalogues, Germany.)

2. Rotor $A^\backslash$ (cup, pot, and camera) with its rotor spindle A with a diameter of φ9 mm.

3. Pressing pulley (tension roller) that presses the rotor spindle against the elastic surfaces of Twin Disc rollers to secure continuous contact between the rotor spindle and the polyurethane cover of the Twin Disc rollers either on the front or the back sides.

4. Tangential belt R covers the full side of the rotor spinning machine and rotates the wharves W of the Twin Disc rollers' shafts (spindles) and consequently the Twin Disc roller rotates with a certain RPM due to the wharves' diameters and the linear speed of the belt V_R. The rotation of the rollers of the Twin Disc drive will rotate rotor spindle shaft A with multiplying speed and folds of the rollers, that is, if the Twin Disc rollers rotate at 12.5 kRPM, the rotor $A^\backslash$ and its spindle A will run at 100 kRPM (ratio of diameters $72/9 = 8$).

The kinematic analysis of the multiplying drive is shown in Figure 3.4 where the horizontal axis represents the rotor spindle accelerating time in centiseconds (CS) and the vertical axis represents the linear speed in kilometers per hour (km/h).

When it begins running, the polyurethane rings (elastic [polymeric] coverings) reach their running speed $V_v = 25$ km/h in a time of 60 CS; this is represented by the first curve (left-side curve). For the wharves of the Twin Discs' roller, they reach their running speed V_ω in 65.0 s (third curve from left side) as the belt's linear speed $V_R = 6$ km/h (third curve from left side). If we assume that a rotor diameter of φ60 mm runs at 90 kRPM, then its surface speed will be 170 km/h. This speed must be attained in 350 CS.

The rotor's inertia and air resistance is shown by the second curve from the left side. The surface speed of rotor spindle A in such a case is 42 m/s and is the same surface speed of the Twin Disc that will be reached in 3.5 s (as rotor $A^\backslash$) (second curve from left side of V_S in Figure 3.4). If we compare the two curves V_E (first curve on the left) and V_S (middle curve), a large dashed area between them is due to the slip percentage between rotor shaft A and the elastic (polymeric coverings).

This slip percentage will create heat that will affect the properties and performance of the rings of polyurethane and, consequently, the rotor speed will be affected where yarn breakage is increasing. The relative merits of the multiplying drive are

1. There is no sufficient guarantee of a constant RPM of the rotor.

2. The polyurethane rings must be replaced due to wear from time to time, which will affect the rotor's rotation speed.

3. The tangential belt requires high pressing force that will impact negatively on the operation of the Twin Disc.

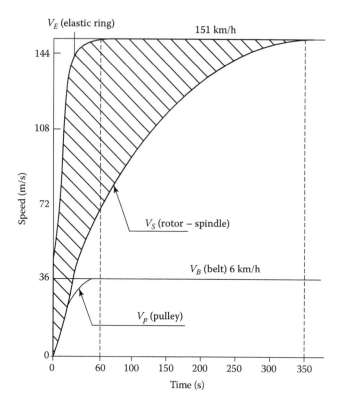

FIGURE 3.4

Kinematic analysis of the multiplying drive of the rotor. V_E, curve of elastic covering speed; V_S, curve of rotor spindle speed; V_p, smallest curve of rotor wharve; V_B, horizontal belt = 47 m/s. (From Sussen Group, ITMA, 1971, Rotor speeds up to 80.000 rpm, Special Catalogues, Germany.)

For all of these reasons or drawbacks, a second-generation rotor has been designed.

In the multiplying rotor drive, it must be noted that the drive starts from a low speed (V_R = 10 m/s, i.e., V_v = 42 m/s—12.5 kRPM) and moves to a high speed (rotor speed) of 90 kRPM. This means that the speed multiplies several times.

B. *The reducing drive of the second-generation rotor*

Figure 3.5 shows the details of the indirect reducing rotor drive (second generation of the Sussen Twin Disc drive). This drive is very similar in design to the multiplying drive (see Figure 3.3). The main difference between the two drives is the location of the tangential belt. In the reducing drive, the tangential belt comes into direct contact with the rotor spindle A, that is, it starts from the high linear speed of the belt = 47 m/s and moves toward the low speed of the Twin Disc rollers.

This means that a reduction in the speed is achievable.

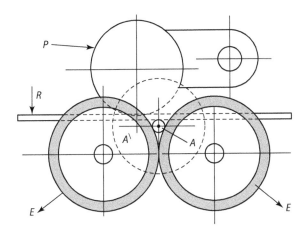

FIGURE 3.5
Kinematic analysis of rotor's drive. P, pressing roller (tension roller); R, tangential belt; A, rotor spindle cross section (φ9 mm); E, elastic polymeric ring (φ72 mm, 12.5 kRPM); A⟍, rotor (100 kRPM). (From Sussen Group, ITMA, 1971, Rotor speeds up to 80.000 rpm, Special Catalogues, Germany.)

The kinematical analysis of the reducing rotor drive is shown in Figure 3.6. As the tangential belt causes the rotational speed of the rotor directly, then we can expect that the tangential belt's linear speed V_R will be high and equal to 47 m/s. Then, the surface speed of the rotor spindle V_A will be the same as the surface speed of the Twin Disc rollers V_v. This will make the accelerating time 3.5 s (the rotor $A⟍$ and the rotor spindle A are one block).

The accelerating curves V_A (spindle + rotor) and V_v (Vulkollan rings) are too close to each other; therefore, the shaded area between them is so small that low heat is generated. Consequently, the deterioration of the polyurethane rings leaves them too weak to provide more rotor speed stability.

The space between the tangential belt V_R and the rotor spindle shaft A is relatively large and the greater slip percentage and heat generation mean that this heat can be absorbed and dissipated easily via the metallic surfaces of rotor spindle A and rotor $A⟍$.

A great advantage of the reducing drive is that the tangential belt thickness is reduced, which will decrease the bending stress caused by repeated bending between the tension rollers and rotor spindles and increase the belt's lifespan. This can be explained (the low thickness of the belt) by the fact that the power requirement for the rotor is relatively constant (irrespective of the type of drive), that is, the belt tension multiplied by its linear speed is constant (power). When the belt speed is increased to 47 m/s, the belt tension will consequently be reduced, that is, the belt cross section will be low.

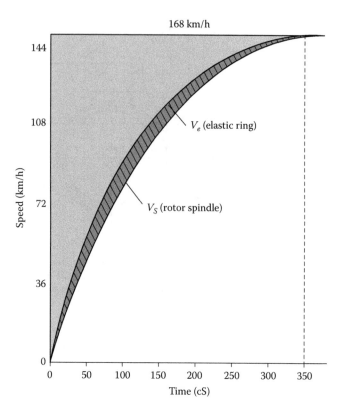

FIGURE 3.6
Kinematical analysis of a rotor drive-reducing drive. V_e, first upper curve: elastic ring covering speed; V_S, second curve: rotor spindle speed. (From Sussen Group, ITMA, 1971, Rotor speeds up to 80.000 rpm, Special Catalogues, Germany.)

As shown previously, the drive moves from a high speed (rotor) to a low speed (Twin Disc rollers), that is, the speed is reduced.

3.5 Summary Points

1. The rotor of the rotor spinning machine can be driven via three systems; the central group drive, the multiplying drive, which has fallen into disuse, and the reducing drive, which is widely applied in commercial rotor spinning machines.

2. The cheapest drive is the central group drive, but it is noisy due to the many rolling and moving elements, as well as the low running speed of the rotor.

3. The reducing drive has many relative merits such as the high speed of the rotor and lower noise levels due to the elimination of the classic

single-row radial rotor spindle bearing. It is also, to a certain degree, environmentally friendly. In addition, the tangential belt lifespan is increased.

4. For any belt drive, the tension in the tight side is always higher than the tension in the slack side by the ratio of $e^{\mu\theta}$, where e-const. = 2.71, μ is the friction coefficient between the belt and the other contacted surface, and θ is the central angle of contact between the belt and its contacted surface.

5. The stress on the belt is divided into three types: stress due to tension in the tight side (σ_1), stress from centrifugal tension (σ_v), and stress due to the belt bending around the wharve shaft and pulleys.

6. The width of the tangential belt is restricted due to different belt stresses.

7. The belt thickness in the reducing rotor in the direct rotor drive is reduced due to the belt speed increasing from 144 to 169 km/h (18%).

8. In modern rotor spinning machines, there are two problems: (1) the turbine size has reduced over two decades and has certainly been connected with rising power requirements at ever-high speeds; and (2) the temperature of air leaving the turbine (rotor) is very high (Lord, 2012).

9. The rotor speed in RPM, depending on the rotor's diameter, can be calculated by using a Shlafhorst-modified formula (Shlafhorst catalogue, private communication).

$$N_R = \frac{420 \times (E+6)}{D_R} (\text{RPM}) \tag{3.9}$$

where:
N_R = Rotor's speed in RPM
D_R = Rotor's diameter in millimeters and $4.20 \times (E + 6)$ – constant

Equation 3.9 is based on a rotor's surface speed = 220 m/s (792 km/h).

Review Questions

Q#1: Describe in detail the design of one of the following drives: the central group drive, the reducing direct rotor drive, or the individual drive of rotor spinning machines.

Q#2: Why has the multiplying indirect rotor drive of a rotor spinning machine fallen into complete disuse?

Q#3: What are the different stresses that are generated in the tangential belt drive of a rotor spinning machine?

Q#4: Referring to the kinematical analysis for both the multiplying and reducing drive, the dashed area between the accelerating curves of the elastic covering (ring) and the rotor spindle shaft is greater for the multiplying drive than the reduced drive. Why?

Answers to Review Questions

> Q#1: For the individual drive, see Section 3.2; for the center group drive, see Section 3.3; and for the reducing indirect rotor drive, see subtopic B in Section 3.3.
>
> Q#2: The large dashed area between the acceleration curves of the elastic ring covering V_e and the rotor spindle curve V_A means that the belt slip percentage is greater and consequently more heat is generated that will affect the polyurethane rings around the rollers of the Twin Disc. This will affect the stability of the rotor speed.
>
> Q#3: See Formulae 3.3 through 3.7.
>
> Q#4: Due to a greater slip percentage between the elastic ring covering, V_e accelerating curves, and the rotor spindle V_s accelerating curve (see Figure 3.4).

Review Problems

> Q#1: In a central group drive, for a rotor's wharve $\varphi = 18$ mm and linear belt speed $V_R = 10$ m/s. What is the rotor's RPM?
>
> Q#2: Repeat Problem 1 for linear belt speed where $V_R = 40$ and 47 m/s. Keep the other factors constant.
>
> Q#3: For the reducing indirect rotor drive, the belt speed is 47 m/s and the rotor spindle is $\varphi = 9$ mm. What will be the rotor's RPM?
>
> Q#4: In Problem 3, what will be the rotor's RPM? (Apply the Shlafhorst formula.)
>
> Q#5: Calculate the rotor's surface speed. Use the data from Problem 4.

Answer to Review Problems

> Q#1:

$$V = \pi dN$$

$$10 \times 60 = \frac{\pi \times 18}{1000} \times N$$

$$\therefore N = 10.616 \text{ RPM}$$

> Q#2: By proportionality

$$N_1 = 4 \times 10,616 \quad (40/10)$$

$$= 42,464 \text{ RPM}$$

$$N_2 = 4.7 \times 10,616 \quad (47/10)$$

$$= 49,895\,\text{RPM}$$

Q#3:

$$V = \pi dN$$

$$47 \times 60 = \frac{\pi \times 9}{1000} \times N$$

$$\therefore N = 99,768\,\text{RPM}$$

$$\cong 100\ \text{kRPM}$$

Q#4: Shlafhorst formula:

$$N_R = \frac{3.62 \times 10^6}{D_R}$$

$$\therefore 100,000 = \frac{3.62 \times 10^6}{D_R}$$

$$\therefore D_R = 36\ \text{mm}$$

Q#5:

$$V_R = \pi D_R N_R / 60$$

$$= \pi \times 0.036 \times 100,000$$

$$= 188\,\text{m/s}$$

Bibliography

Elha/wary I.A., 2013, Mechanics of the rotor spinning machine, Lecture notes, TED, Alexandria University, Alexandria, Egypt.

Rengasamy R.S., 2002, *Mechanics of Spinning Machinery*, NECUTE, Delhi, India.

Sentil Kumar R., 2015, *Process Management in Spinning*, CRC Press Taylor & Francis Group, New York.

Shlafhorst, 1980, Printed materials (private communications).

SKF Group, 2013, Roller Bearings Catalogue, Sweden.

References

Lawrence C.A. and Chen K.Z., 1984, Rotor spinning, *Textile Progress*, 13/4, 54–58.
Lord P.R., 2012, *Hard Book of Yarn Production*, WP Publisher, Cambridge, UK.
Moscow State University of Design & Technology, Moscow, RFU, 1978.
Sussen Group, ITMA, 1971, Rotor speeds up to 80.000 rpm, Special Catalogues, Germany.

4

Rotor Power Requirements

4.1 Introduction

The power requirements for any rotor spinning machine are divided into two parts:

1. Mechanical power: This is required to overcome the frictional resistance inside the antifriction or sleeve bearings. These frictional resistances are created due to dynamic reactions in the bearings and consequently due to static unbalance. They can be minimized through the dynamic balancing of the rotor.

2. Air resistance around the rotor: The power requirements, due to either bearing friction or air resistance, are dependent on the value of the rotor's eccentricity and the amount of air around the cup. This can be explained as follows:

   a. As quality control of rotor eccentricity after manufacturing requires that it must not exceed more than 1 μm (micrometer), the frictional resistance in the bearings of the rotor spindle is neglected. It was found experimentally that the bearing friction requires 0 W for a rotor speed of 40 kRPM; the bearing friction usually accounts for 8%–10% of the total power.

   b. The air resistance around the rotor causes an additional power requirement, but its value depends on the amount of air around the rotor. As the air quantity around the rotor decreases, the required watts per rotor will decrease. Of course, the total power of the rotor will be the summation of (a) and (b).

Active example 1
Calculate the rotor's power requirements due to bearing friction by using only the following data:

- Rotor diameter (φ) = 56 mm
- Rotor mass = 70 g

- Rotor mass eccentricity $e \leq 1$ μm
- Rotor revolutions per minute (RPM) = 30,000
- Bearing coefficient of friction $\mu = 0.00'$
- Rotor spindle $\varphi = 9$ mm

Solution
The centrifugal force (cN) on the rotor is

$$cN = M\omega^2 e$$

$$= 0.07 \times \left[30,000 \times \frac{2\pi}{60} \right]^2 \times 0.001$$

$$\therefore \ cN = 690 \ N$$

Friction force $F = \mu \times 690$

$$= 0.001 \times 690$$

$$\cong 0.690 \ N$$

Frictional torque $T = F \times r$

$$= 0.690 \times \frac{9}{2000}$$

$$= 3.1 \times 10^{-3} \ Nm$$

Power $= T \times \omega$

$$= 3 \times 10^{-3} \times 3140$$

$$= 9.75 \ W$$

$$\cong 10 \ W$$

If it is assumed that the rotor's power requirement is 150 W, then the bearing friction power is 6.67% ($\cong$7%).

In general, the rotor's power requirements are mainly determined by air drag around the rotor due to (a) the small bearing frictional power; (b) the rotor's high speed; and (c) the dimensions of the rotor itself.

4.2 Methods of Power Calculation

There are three methods of rotor power calculation.

4.2.1 The Shirley Institute (1968)

The Shirley Institute (1968) proposed two formulae:

The power of the rotor without considering its driving unit. The power is calculated by

$$P = 3.5 \times 10^{-12} \times N^{2.5} \times D^{3.8} \text{W} \tag{4.1}$$

where:
P = Power requirements of the rotor
N = Rotor's RPM
D = Rotor's maximum diameter in inches

The restrictions of Formula 4.1 are as follows:

1. It is not applicable when

$$N \times D^2 < 60.000 \tag{4.2}$$

2. It is not applicable when $D \leq 2.5$ in. (63.5 mm) and also when $D > 5$ in. (127 mm), that is, when

$$2.5'' > D > 5''$$
$$63.5\,\text{mm} > D > 127\,\text{mm} \tag{4.2a}$$

Formula 4.1 considers the rotor to be an inherently thin disc, but the rotor actually has a thickness and height. Formula 4.3 gives the rotor's power taking into consideration the driving unit's power requirements:

$$P = 8.2 \times 10^{-12} \times N^{2.5} \times D^{3.8} \tag{4.3}$$

Using the equation items in Formula 4.1 in comparison to Equations 4.2 and 4.3, we find that the driving efficiency η is 42.7% (Shirley Institute, 1968).

$$\eta = \frac{3.5}{8.2} \times 100$$

$$= 42.7\%$$

This means that if the rotor's total power is 100 W, then around 43 W of this power goes to the drive and the rotor itself requires only 57 W.

Active example 2
Using the following data, calculate the power required to drive a rotor unit with and without a drive:

- Maximum rotor diameter $\varphi = 63.5$ mm (2.5″)
- Rotor speed $= 30$ kRPM

Solution
To fulfill the conditions of Formula 4.1:

a. $N.D^2 = 30,000 \times (2.5)^2$

$\qquad = 187,500 > 6,000$

b. $D = 2.5″$

The rotor's power with or without a drive is not applicable.

4.2.2 Moscow State University of Design & Technology, msta.ac

This method calculates the rotor's power requirements without considering the driving unit. The rotor is divided into three parts: the disc, cylinder, and cone parts. The concept here is the aerodynamic resistance of the rotor (see Figure 4.1).

1. *Disc part* (P_d)

   This part represents the base of the rotor with the maximum rotor diameter, where it is considered to be an infinitesimally thin disc. The power of part P_d is

$$P_d = \frac{\pi}{5} c_{fd} . \rho \omega^3 R^5 \tag{4.4}$$

where:

P_d = Power requirements of rotor (disc part)

$\pi = 3.14$

ρ = Air density = 1.18 kg/m³

R = Maximum rotor radius

ω = Rotor's radius speed

C_{fd} = Air coefficient (turbulent flow), calculated using the formula:

$$C_{fd} = 0.057 / \sqrt[5]{R_e} \qquad (4.5)$$

where R_e is the Reynolds number, which is determined by the equation:

$$R_e = \frac{V.R}{\gamma} \qquad (4.6)$$

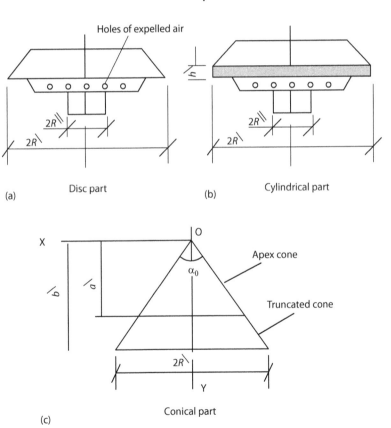

(a) Disc part

(b) Cylindrical part

Holes of expelled air

(c) Conical part

FIGURE 4.1

Geometrical rotor parts. (From Moscow State University of Design & Technology, Moscow, RFU, 1974.)

where:

V = Circumferential speed of the rotor

R = Maximum rotor radius

γ = Kinematic air viscosity = 1.57×10^{-5} m²/s (see Figure 4.1a)

2. *Cylindrical part*

The power required for a rotor's cylindrical part with height h is calculated using Dorfman's formula:

$$P_c = \pi c_{fc} \rho \omega^3 R^5 h \tag{4.7}$$

where:

P_c = Cylindrical part of the rotor's power requirements

c_{fc} = Air drag coefficient between air and cylindrical part of the rotor

ω = Rotor's radian speed

h = Height of cylindrical part

R = Maximum radius of the cylindrical rotor part

The value of c_{fc} is calculated using Basalev's equation:

$$\frac{1}{\sqrt{c_{fc}}} = -0.6 + 4.07 \log\left[R_e \sqrt{c_{fc}} \right] \tag{4.8}$$

or c_{fc} can be estimated using the formula:

$$c_{fc} = 0.096 c_{fd} \tag{4.9}$$

See Figure 4.1c.

3. *Conical part (P_k)*

The power required for the conical part of the rotor, P_k, is determined by the formula:

$$P_k = B c_{fk} . \rho . \omega^3 [b^5 - a^5] \tag{4.10}$$

where b is a constant that is calculated using the formula:

$$B = \frac{\pi}{32} \cdot \frac{k^2}{5} \cdot \sqrt{4+k^2} \qquad (4.11a)$$

where:

$$k = \frac{D}{b}$$

D = Maximum rotor diameter

b = Total cone height (Figure 4.1c)

The apex half angle α is calculated by

$$\left. \begin{array}{c} \tan\left(\dfrac{\alpha_0}{2}\right) = \dfrac{R}{b} = \dfrac{k}{2} \\[2mm] \therefore k = 2R/b \end{array} \right\} \qquad (4.11b)$$

where ∂' is the distance from cone apex (0) to the tip of the truncated cone at the rotor's lower conical part (Figure 4.1c)

The total power requirement using the Moscow method is the summation of Equations 4.4, 4.7, and 4.10, that is,

$$\text{Total rotor power} = \frac{\pi}{5} \cdot c_{fd} \cdot \rho \cdot \omega^3 \cdot R^5 + \pi \cdot c_{fc} \cdot \rho \cdot \omega^3 Rh + Bc_{fk}\rho\omega^3\left[b^5 - a^5\right]$$

$$P_{total} = \rho\omega^3\left[\frac{\pi}{5} \cdot c_{fd} \cdot R^5 + \pi \cdot c_{fc} \cdot R\grave{h} + Bc_{fk}\left[b^5 - a^5\right]\right] \qquad (4.12)$$

Active example 3

Use the same data in Active Examples 1 and 2 for the rotors of a rotor spinning Machine A.

Solution

By applying the previous data for the rotor's conical part, the results shown in Table 4.1 can be obtained. The total rotor power using the Moscow method is the summation of power for all three parts of the rotor (Formula 4.12).

The first column in Table 4.1 gives the measured results using a Siemens wattmeter. As a rule, the control of calculated values is the measured values;

TABLE 4.1

Theoretical and Experimental Results of Rotor Power (Watts) of Machine A

	Power in Watts by Formulae without Drive		Rotor
Experimental Results	**Shirley Inst.**	**Moscow (MSAT)**	**kRPM**
39.3 W	17.74	38.0	30
	26.08	60.3	35
	48.89	128.1	45
	63.62	175.8	50

the Moscow method is the nearest to the measured results with a rotor speed of 30 kRPM, that is, it neglects the other parts of the rotor (cylindrical and conical). The percentage error is 2.7% for the Moscow method while the percentage error for Shirley is around 55%. Thus, the Shirley method is not suitable.

Table 4.2 shows the rotor's power calculated using the Shirley method with the rotor's drive and the measured values from a Siemens wattmeter for Machine B.

From Table 4.1, it is shown that the difference is greater between watts in column 2 and 3. This is due to the small rotor diameter of 1.77" (45 mm), which is less than the Shirley Institute's minimum rotor diameter of 2.5", and the high rotor speed.

It is important to note the following:

1. According to the Shirley Institute, the Reynolds number R_e for a rotor is calculated by

$$R_e = 1.4 \times N \times D^2 \tag{4.13}$$

  where N and D are the rotor's RPM and diameter, respectively.

2. The critical Reynolds number that separates the laminar air flow and the turbulent air flow is given by the value of R_{ec}, where

TABLE 4.2

Experimental and Theoretical Results of Rotor Power (Watts) of Machine B

			Rotor Power without Drive	
Rotor RPM	**Siemens Wattmeter Results (W)**	**Shirley Method with Drive**	**Shirley**	**Moscow**
30,000	16.77	17.27	7.2	7.5
35,000	22.82	25.18	10.5	9.1
45,000	37.72	47.47	19.7	25.2
50,000	46.58	61.39	25.6	34.5

$$R_{ec} = 7 \times 10^5 \,(700,000) \tag{4.14}$$

3. The opening device of a rotor spinning machine requires 22 W of energy for an opening cylinder speed of 8 kRPM.

4.2.3 Zurich Method (SFIT)

Zurich, the method of the Swiss Federal Institute of Technology (SFIT), introduced to the rotor spinning machine a group of empirical formulae, one of which concerns the rotor as shown in Equation 4.15 (rotor with driving unit):

$$P = 36 \times \dot{d}_R \times \dot{d}_w \times n_R'^2 \tag{4.15}$$

where:
P = Required power to drive the rotor with a driving unit
$\dot{d}_R$ = Rotor diameter in meters and rotor speed in kRPM divided by 1000 or the rotor's RPM, that is, without kilos (k)
$\dot{d}_w$ = Wharve diameter in meters (Krause and Soliman, 1980)

Active example 4
   For rotor diameters 67, 95, and 45 mm with wharve diameters of 18, 18, and 11.5 mm, respectively, and with a rotor speed of 50 kRPM, calculate the corresponding power for each rotor.

Solution
   Formula 4.15 is applied and the results are as follows:

1. For rotor φ67 mm and wharve φ18 mm, then $P = 109.0$ W
2. For rotor φ45 mm and wharve φ18 mm, then $P = 59.0$ W
3. For rotor φ45 mm and wharve φ11.5 mm, then $P = 46.0$ W

___

4.3 Summary Points

1. The power requirements of the rotor can be calculated using the Shirley, Moscow, and Zurich methods. The Moscow method does not consider the effect of the drive on the rotor's power and depends mainly on the air drag. The other methods take into consideration both the bearing friction and the air friction with the rotor's body.
2. The Reynolds number is vital for differentiating between laminar air flow and turbulent air flow. Also, it is important to take into account when calculating the air drag coefficient with the rotor's surface.

3. The rotor's power measured by a wattmeter is a required reference to check the theoretical values calculated.

4. Usually, the empirical formulae are vital and active within the range of their experimental work. In other words, their validity is restricted.

5. The measured rotor's power is 39 W for rotor diameter φ63.5 mm and speed 30 kRPM, while for rotor diameter φ45 mm and speed 30 kRPM, it is 17 W. The measured values of a rotor's power take into consideration the rotor's driving unit.

6. For a rotor's power, the Reynolds number is

$$R_e = \frac{V.R}{\gamma}$$
(4.16)

where:
V = Surface speed of the rotor (m/s)
R_e = Rotor's radius
γ = Kinematic viscosity of the air (=1.57 × 10^{-5} m^2/s)

7. The critical Reynolds number $R_{ec} = 7 \times 10^5$

8. The definition of the Reynolds number for the Shirley method is

$$R_e = 1.4 \times N \times D^2$$
(4.17)

where N and D are the rotor's RPM and diameter in inches, respectively.

9. For the rotor spinning machine, as a whole or its movable parts, power is measured by a wattmeter; there are two main types of wattmeter in general use: (a) the dynamometer or electrodynamic type; and (b) the induction type (Theraja, 1979).

10. Energy = 9.81 J (joule)

11. Power = 1 Ps (horsepower)

= 736 W

= 1.0 kcal/h (kilocalories per hour)

= 1.163 W

= 1 hp

= 746 W

12. Kilowatts $= 1.36$ Ps

$$= 102 \, mkg/s$$

$$= 0.236 \, kcal/s$$

13. Ps $= 75 \, mkg/s$

$$= 0.176 \, kcal/s$$

14. Dynamic viscosity $= 1$ poise

$$= 0.1 \, pas$$

15. Kinematical viscosity $= 15 \, t$

$$= 1 \times 10^{-4} \, m^2/s$$

$$= 1 \, cst$$

$$= 1 \times 10^{-6} \, m^2/s$$

16. British thermal unit (BTh) $= 0.252 \, kcal$

17. Pound $-$ force $\left(1 \, lbf\right) = 4.4482 \, N$

18. Psi $\left(pounds \, / \, sq \, inch\right) = 6.8948 \, 10^3 \, pa$

19. One lbf $-$ in $= 0.113 \, Nm$

20. One ft $-$ lbf $/s = 1.3558 \, W$

21. The maximum allowable rotor rotational speed (n max) in the rotor spinning machine is determined by the formula (Sentil Kumar, 2015):

$$n \, max = \frac{2700}{'D_R} \times \sqrt{\frac{T_{rsy}}{R}}$$

where:
n max $=$ Maximum allowable RPM of the rotor
$'D_R =$ Rotor diameter (m)
$T_{rsy} =$ Tenacity of rotor-spun yarn
$R =$ Ratio between theoretical and actual draw of tensions

Review Questions

> *Q#1:* Mention the different formulae for calculating the rotor's power in a rotor spinning machine.
>
> *Q#2:* Compare qualitatively the different techniques for calculating a rotor's power.
>
> *Q#3:* Define and explain the Reynolds number. What is the value of the critical Reynolds number?
>
> *Q#4:* Mention the restrictions of each of the different applied techniques for calculating the rotor's power.

Answers to Review Questions

> *Q#1:* See Formulae 4.1, 4.3, 4.12, and 4.15.
>
> *Q#2:* The Shirley method considers the rotor to be an infinitesimally thin disc, but the rotor base (at maximum diameter) actually has a thickness (h).
>
> The Academy of Textiles at Moscow considers the rotor to be composed of three parts—the disc, cylinder, and cone—and it is subjected to air drag.
>
> The Zurich method depends on the empirical formulae that depend on the experimental results. Also, it introduces the rotor-spindle wharve as an effective element.
>
> *Q#3:* The Reynolds number is a dimensionless figure (value) that is applied mainly in aerodynamics for measuring the movable masses in the air as coefficients of air drag and differentiating between laminar and turbulent air flow. The definitions of R_e are

$$R_e = \frac{v.R}{\gamma} \text{(MSAT)}$$

$$= 1.4 \times N \times D^2 \left(\text{Shirley value}\right)$$

where:
> v = Rotor's surface speed R
> D = Rotor's radius and diameter, respectively
> N = Rotor's RPM
> γ = Kinematic viscosity

> *Q#4:* The different restrictions are as follows:
>   a. Shirley method
>   1. The disc diameter cannot be less than 2.5" (63 mm).
>   2. The product $N \times D^2$ cannot be less than 60,000.

b. Moscow State University of Design & Technology method

The restriction is connected with the accuracy of the air drag coefficient with the disc part c_{fd} and both the cylindrical and conical parts $c_{fd}=c_{fk}$.

c. Zurich method

The restriction is related to the boundary conditions of the experimental work, for example, rotor spindle wharves' diameters of 11.5 and 18 mm and rotor's diameters of 45 and 67 mm. If Equation 4.15 is applied over and below these values, the results will contain certain approximations.

Review Problem

Q#1: A rotor spinning machine has rotors with the following properties:

- Rotor diameter = 35 mm
- Rotor speed = 120 kRPM
- Height of its base $h = 10$ mm
- Nose cone apex angle = 54°

Guidelines:

Assume any of the required data; compare the rotor's power by using the different methods of calculations.

Answers to Review Problem

A. Shirley method

The rotor φ33 mm = 1.299", that is, it is less than 2.5". Thus, the method is restricted. The product $ND^2 = (120.000) \times (1.299)^2 = 219{,}362 > 6{,}000$.

Again, the Shirley method is not applicable.

B. Moscow State University of Design & Technology method

It is preferable to determine or calculate in advance some items of the method as will be shown later. We then calculate the power for each part (disc, cylinder, and conical cup) and separately make summations to calculate the total power using Equation 4.12 as shown next:

From the solution using the MSAT method

It will be shown later that the conical part plays the main role in power calculation.

Reynolds number

$$R_e = \frac{V.R}{\gamma}$$

$$V = \pi dN/60 = \pi \times 0.033 \times 120.000/60$$

$$= 207.24 \text{ m/s}$$

$$R = 0.0165 \text{ m}$$

$$\gamma = 1.57 \times 10^{-5} \text{m}^2/\text{s}$$

$$\therefore R_e = 2.178 \times 10^5$$

By using the Shirley formula,

$$R_e = 1.4 \times N \times D^2$$

$$= 1.4 \times 120.000 \times \left(\frac{33}{25.4}\right)^2$$

$$= 2.026 \times 10^5$$

The R_e of MSAT is larger than the R_e of Shirley by 7.5%. We apply the smaller Reynolds number value.

Coefficient of friction with disc c_{fd}

$$c_{fd} = 0.057/\sqrt[3]{R_e}$$

$$= 0.057/\sqrt[5]{2.178 \times 10^5}$$

$$= 2.4427 \times 10^{-5}$$

Coefficient of friction of air with cylinder c_{fc}
According to Formula 4.9:

$$c_{fc} = 0.097 \times c_{fd}$$

$$= 2.3695 \times 10^{-6}$$

Air drag with cone

$$c_{fk} = c_{fd}$$

$$= 2.4427 \times 10^{-5}$$

Cone conicity k

Nose cone apex angle $\alpha = 54°$

$$\therefore \frac{\alpha}{2} = 27$$

$$\therefore \tan\frac{\alpha}{2} = 0.5095 = \frac{k}{2}$$

$$\therefore k = 1.01591 = \frac{D}{b} = \frac{33}{b}$$

$$\therefore b = 32.4\,\text{mm}$$

Assume $a = 0.7$ and $b = 22.7$:

$$\therefore k^2 = 1.0321$$

$$\therefore B = 0.0454$$

$$b^5 - a^5 = 0.0324^5 - 0.0227^2$$

$$= 3.5705 \times 10^{-8} - 6.0274 \times 10^{-9}$$

1. *Disc part*

$$P_d = \frac{\pi}{5} \times 2.4427 \times 10^{-5} \times 1.18 \times 12560^3 \times 0.0165^5$$

$$= 0.628 \times 2.4427 \times 10^{-5} \times 1.18 \times 1.9813 \times 10^{12} \times 1.2230 \times 10^{-9}$$

$$= 4.38620 \times 10^{-2}$$

$$= 0.0439\ \text{W}$$

2. Cylindrical part

$$P_c = \pi \times 2.3695 \times 10^{-6} \times 1.18 \times 12560^3 \times 0.0165^4 \times 0.01$$

$$= 7.44 \times 10^{-6} \times 1.18 \times 1.9814 \times 10^{12} \times 7.412 \times 10^{-8} \times 0.01$$

$$= 10^{-2} \times 1.2893$$

$$= 0.012893 \text{ W}$$

3. Conical part

$$P_k = 0.0454 \times 2.4427 \times 10^{-3} \times 1.18 \times 12560^3 \times \left[2.9677 \times 10^{-8} \right] = 7.6948 \times 10^{11}$$

$$= 7.6948 \text{ W}$$

$$\cong 8 \text{ W}$$

$$= 0.0439 + 0.0129 + 8$$

$$\therefore P_{total} = 8.057$$

$$\cong 8 \text{ W} \left(\text{without drive} \right)$$

$$\cong 16 \text{ W} \left(\text{driving } \eta = 50\% \right)$$

C. Zurich technique (SFIT)

$$P = 36 \times d_R' \times d_\omega' \times n_R'^2$$

$$= 36 \times 0.033 \times 0.018 \times 120^2 \text{ (wharves } \varphi \text{ mm)}$$

$$= 25.66$$

$$\cong 26 \text{ W}$$

Seemingly, it is the Zurich method that gives the most satisfactory value. From our point of view, the Zurich method is the closest in practice, and has the least restrictions of all the methods. The [msta.ac] method with drive value is only 62% of the Zurich method value.

Bibliography

Moscow State University of Design & Technology (msta.ac), 1975, Rotor's machines, Notes Lab, Moscow, RFU.

References

Krause E. and Soliman H.A., 1980, The power distribution in a rotor spinning machine, Private communication, Swiss Federal Institute of Technology (SFIT), Switzerland.

Moscow State University of Design & Technology, Moscow, RFU, 1974.

Sentil Kumar R., 2015, *Process Management in Spinning*, CRC Press, Taylor & Francis Group, New York.

Shirley Institute, 1968, Break spinning: The final report of a three years investigation undertaken by the Cotton, Silk and Man-Made Fibres Research Association, Manchester, UK.

Theraja B.L., 1979, *Electrical Technology*, 17th revised edition, S. Chand & Company Ltd., New Delhi, India.

5

Power Distribution in the Rotor Spinning Machine

5.1 Introduction

In a rotor spinning machine, the basic elements are the feed, opening mechanism, rotor, delivery roller, winding drum, ventilation fan, and control system.

During running time, the energy cost of producing 1 kg of yarn is about 6%–12% of its total cost; the spinning process itself consumes about 60%–80% of the total spinning mill energy.

5.2 Opening Mechanism

The opening mechanism consists mainly of an opening cylinder and combing roll (toothed drafting). The power of the opening mechanism can be divided into the technological toothed drafting of the fed sliver and the mechanical power of the combing roller.

5.2.1 Technological Opening Power

The technological opening power of the fed sliver to the feeding mechanism is determined by different formulae depending on the type of opened material (toothed drafted material) such as noils of a cotton comber, virgin cotton (medium-large), staple cotton, or artificial fibers.

a. For a *cotton comber noils* opening, the power is

$$P = 0.013 \times F \times n_0^2 \, \text{W} \tag{5.1}$$

where:
  P = Opening power in watts
  F = Rate of fiber feed in grams per minute = feed speed × sliver
     ktex; usually, it ranges from 1 to 8 g/min
  n'_0 = Combing roller revolutions per minute (RPM) divided by 1000; for
     example, if the opening cylinder's RPM is 8000, its value will be 8

b. For a *medium-long staple virgin cotton*, the opening power is

$$P = 0.09(F+1.5) \times n'^{1.5}_0 \ W \tag{5.2}$$

where P is the medium-long staple cotton opening power in watts,
and F and n'_0 are as defined in Formula 5.1.

c. For *artificial fibers*, the opening power is

$$P = 0.09(F+1.5) \times n'^{2.6}_0 \ W \tag{5.3}$$

where P is the artificial fibers' opening power in watts, and F and n'_0
are as defined previously.

5.2.2 Combing Roller Opening

The power required to overcome toothed drafting force, roll eccentricity, and
air drag by the combing roller (mechanical power) is

$$P = S_0 \times d'_w \times n'_0 \ W \tag{5.4}$$

where:
  P   = Mechanical power of the combing roller opening roller in watts
  d'_w = Combing roll spindle wharves in meters
  n'_0 = As defined in Equation 5.1

It is important to note the following:

1. The opening device consists of a feed roll, trumpet, and combing
   roll. The feed roll power is neglected. Therefore, the opening device
   power is mainly due to the combing roll.
2. The power of the combing roll only, that is, without the sliver, is
   referred to as *no-load* power; if we add the technological opening
   power to it, it will be referred to as *full-load* power.

Active example 1
   A rotor spinning machine is fed a cotton sliver (virgin cotton) with a ktex
value of 4 (4 g/m) and a combing roller speed of 8 kRPM. Calculate the

technological opening power, the mechanical power of the combing roll, the no-load power, and the full-load power. Assume any of the required data.

Solution

The surface speed of the combing roll is V:

$V = \pi d n_0 = \pi \times 0.665 \times 8000$ m/min, assuming an opening roller diameter of $\varphi 65$ mm.

$$\therefore V = 1632.8 \, \text{m/min}$$

Assuming that the draft between the combing roll and the feed roll = 10,000, then the surface speed of the feed roll = 0.16328 m/min, and the rate of feed $F = 0.16328 \times 4 = 0.65$ g/min.

$$\therefore F = 0.65 \, \text{g/min}$$

For technological opening power P:

$$P = 0.09(F + 1.5) \times n_0^{1.5}$$

$$= 0.09(0.65 + 1.5) \times 8^{1.5} \qquad (5.5)$$

$$= 4.38 \, \text{W}$$

For combing roll mechanical power P:

$$P = 50 \times 0.024 \times 8 \left(d_w = 24 \, \text{min}\right)$$

$$\qquad (5.6)$$

$$= 9.60 \, \text{W}$$

Equation 5.6 gives the no-load power for the opening device. Equations 5.5 and 5.6 show that the full-load power = 13.98 W $\cong$ 14 W. The percentage accounted for by technological power from the total = 31% (4.38/141×00), thus the no-load power percentage is = 69%.

5.3 Twisting Mechanism

The twisting mechanism includes the rotor, the doffing tube, and the take-up roller. The power of the mechanism without material (yarn) gives the no-load power the material power; the no-load power gives the full-load power for the twisting mechanism.

5.3.1 Effect of Yarn on Rotor

There are two effects of yarn on the rotor: the friction between the yarn and the navel of the doffing tube (lower end of the tube) and the air drag on the yarn inside the rotor. To overcome both of these factors, we need a certain power P, calculated by the equation:

$$P = d'^3_R \times n'^2_R \times \sqrt{\text{tex}} \left[0.03 \mu d'_T . \sqrt{\text{tex}} + 0.567 \times d'^2_R \right] W \qquad (5.7)$$

where:

  d'_R = Rotor diameter in meters
  d'_T = Trumpet (navel) maximum diameter (φ22 mm)
  n'_R = Rotor's RPM divided by 1000
  tex = Rotor-spun yarn tex
  μ = Coefficient between the navel and the yarn is 0.25 for a smooth navel and 0.37 for a notched navel

5.3.2 Rotor with Drive (Equation 4.15)

Active example 2
   A rotor with a maximum diameter φ = 45 mm runs with a speed of 70 kRPM to produce 30s yarn. By assuming any of the required data, calculate the power required to overcome yarn friction with the navel and the air drag on the yarn arm inside the rotor. Also, find the no-load and the full-load power of the twisting mechanism.

Solution
   The yarn power P (as in Equation 5.7) is

$$P = 0.045^3 \times 70^2 \times \sqrt{2} \left[0.03 \times 0.25 \times 0.022 \times \sqrt{20} + 0.567 \times 0.045^2 \right] \qquad (5.8)$$

$$= 0.0038 \ W$$

Then, the rotor's power with its drive P is

$$P = 36 \times 0.045 \times 0.018 \times 70^2 \text{ (rotor-spindle wharves φ18 mm)} \qquad (5.9)$$

$$= 142.9 \cong 143 \ W$$

∴The full-load power is the summation of Equations 5.8 and 5.9, which is 143 W. (Omit Equation 5.8. The rotor power with its drive represents 100% of the full-load power.)

5.4 Rotor-Spun Yarn Withdrawal from the Doffing Tube

The delivery rollers (take-up rollers) withdraw the rotor-spun yarn from the yarn tube (doffing tube) against the yarn tension due to the balloon tension or the centrifugal tension on the yarn arm inside the rotor. The required power to overcome this yarn tension P is

$$P = 0.29 \times d'^2_R \times n'^2_R \times \text{tex} \times V_L \times 10^{-4} \text{ W} \tag{5.10}$$

where:

P = Power in watts
d'_R = Rotor diameter in meters
n'_R = Rotor RPM divided by 1000
tex = Rotor-spun yarn tex
V_L = Surface speed of the delivery roller

Yarn production speed $\cong$40–90 m/min

Active example 3
For a rotor with a speed of 50 kRPM and a diameter of φ40 mm that is used to produce yarn N_m = 40 with a twist factor of α_m = 103, find the power required to withdraw the yarn from the doffing tube.

Solution
The yarn twists

$$\text{tpm} = \alpha_m \sqrt{N_m} = 103 \times \sqrt{40} = 651.4$$

so the yarn tpm

$$= \frac{n_R}{V_L} / V_L = \frac{50.000}{V_L}$$

$$\therefore V_L = 76.75 \cong 77 \text{ m/min}$$

Thus, the required power is

$$P = 0.29 \times 0.040^2 \times 50^2 \times 25 \times 77 \times 10^{-4}$$

$$= 0.22 \text{ W} \left(\text{yarn tex } 25\right) \tag{5.11}$$

5.5 Yarn Winding

The power required to build up the rotor-spun yarn on its spool to form a package P is

$$P = 0.02 \times V_L \times \frac{h'}{d'_T} \tag{5.12}$$

where:

P = Power of package building
h' and d'_T = Package's height or length and diameters, respectively, in meters
V_L = Yarn withdrawal speed

Active example 4

Calculate the package winding (building) power for 24 s yarn with a package height of 250 m and a diameter of φ230 mm. Take $V_L = 77$ m/min.

Solution

$$P = 0.02 \times 77 \times \frac{250}{230}$$

$$= 1.67\,\text{W}$$

(5.13)

5.6 Feed, Delivery, and Winding

The required power to rotate the feed roller, delivery roller, and winding drum is

$$P = 0.06 \times V_L\,\text{W}$$

(5.14)

where:
  P = Driving power for the feed, delivery, and winding rolls
  V_L = Delivery (production) speed

Active example 5

Calculate the required power to drive the feed roll, delivery rolls, and winding drum, assuming $V_L = 77$ m/min.

Solution

The power P is

$$P = 0.06 \times 77$$

$$= 4.62\,\text{W}$$

(5.15)

5.7 Ventilation

The rotor spinning machine has a ventilation fan for cooling and air suction in the spinning box. The required power for each rotor from the fan power is

$$P = 1.16 \times' n_R$$

(5.16)

Active example 6

Calculate the power required for each rotor concerning its share of fan power for a rotor with a speed of 70 kRPM.

Solution

The power P is

$$P = 1.16 \times 70$$
$$= 81.2\,W$$

(5.17)

5.8 Control Systems

The control system of the rotor spinning machine consists of start and stop programs and a robotic system of automatic piecing, cleaning, and patrolling. The power of all the elements of the control system can be neglected.

5.9 Specific Energy

Specific energy (SE) is the energy in watts per hour that is required to produce 1 kg of rotor-spun yarn.

For a rotor spinning machine, the following equation is used for calculating the SE:

$$SE = 0.06 \times tex^{-1.346} \times 'd_R^{0.62} \times '\pi_R^{0.52} \times \alpha_{tex}^{0.915}\ kWh/kg \qquad (5.18)$$

where:

SE = Specific energy required to produce 1 kg of rotor-spun yarn
$'d_R$ = Rotor diameter in meters
$'\pi_R$ = Rotor RPM divided by 1000
α_{tex} = Rotor-spun yarn twist factor in tex

The SE required for 1 kg of ring spun yarn is determined by Equation 5.19 (Krause and Soliman, 1980):

$$SE = 106.7 \times tex^{-1.346} \times '\Delta_r^{13.343} \times 'n_s^{0.917} \times \alpha_{tex}^{0.993}\ kWh/kg \qquad (5.19)$$

where:

SE = Specific energy in kilowatt-hours per kilogram (kWh/kg) of ring-spun yarn
tex = Ring-spun yarn in tex
$'D_r$ = Ring diameter φ in meters
$'n_s$ = Ring-spindle speed in RPM divided by 1000
α_{tex} = Ring-spun yarn twist factor in tex system

Active example 7

For a yarn, a cotton count of 30 was produced on a rotor spinning machine and a ring spinning frame using virgin cotton (medium-long staple). Compare the specific energy for both systems.

Rotor spinning machine

The rotor $\varphi = 45$ mm; the rotor speed = 50 kRPM

Ring spinning machine

The ring $\varphi = 40$ mm, the spindle speed is 15 kRPM, and the yarn twist factor $\alpha_e = 3.3$.

Solution

It is well known that

$$\alpha_{tex} = \alpha_e \times 957.5 = 31.6\,\alpha_m \qquad (5.20)$$

α_{tex} of yarn (ring spinning)

$$\alpha_{tex} = 3.3 \times 957.5$$

$$= 3159.75$$

$$= 3156$$

α_{tex} of rotor-spun yarn

$$\alpha_{tex} = 3.3 \times 1.20 \times 957.5$$

$$= 3791.7$$

$\therefore$ For rotor spinning,

$$SE = 0.06 \times 20^{-1.346} \times 0.04^{0.62} \times 50^{0.52} \times 3792^{0.915}$$

$$= 0.272 \text{ kWh/kg}$$

For ring spinning,

$$SE = 106.7 \times 20^{-1.482} \times 0.04^{3.343} \times 15^{0.917} \times 3156^{0.993}$$

$$= 0.149 \text{ kWh/kg}$$

The ratio between the SE of rotor and ring spinning machines is $0.272/0.149 = 1.40$, that is, the SE for a rotor spinning machine to produce 1 kg of 30 s count yarn is 40% higher.

Furthermore, SE is very important when calculating the total cost of yarn production.

It is important to note the following:

Referring to the SE equations for both ring and rotor spinning machines, the following remarks are important:

- For yarn count 10–20 s (30–60 tex), the SE for both machines is equal when assuming the rotor $\varphi = 45$ mm, the speed $= 35$ kRPM, and the twist factor $= 5000$ (tex system). For high rotor speeds from 55 to 75 kRPM, the SE for the rotor is 40% higher than the ring spinning machine.
- For cotton yarn counts equal to or less than 10, the SE for the rotor is too small compared to the ring spinning machine.
- For cotton counts of 20 or more, the SE of the rotor spinning machine is higher than the ring spinning machine.
- SE Equation 5.18 does not consider the power savings due to canceling the flyer frame and canceling the winding machine.
- Equations 5.5 through 5.16 give the motor's output power. On the contrary, Equations 5.18 and 5.19 give the input power to the motor.

5.10 Summary Points

1. The relationships between different twist factors are

$$\left. \begin{aligned} \alpha_{tex} &= 957.5.\alpha_e \\ &= 31.70.\alpha_e \\ \alpha_m &= 30.30\ \alpha_e \end{aligned} \right\} \tag{5.21}$$

where α_{tex}, α_m, and α_e = twist factor in tex, metric, and English (cotton) systems. The twist factor of a rotor yarn is higher than that of a ring yarn by about 15%–20%.

2. The relationship between the different numbering systems of yarn are

$$\left. \begin{aligned} \text{Nec}\ (\text{Ne}) \times \text{tex} &= 590.5 \\ \text{Nec}\ (\text{Ne})\ /N_m &= 0.59 \\ N_m \times \text{tex} &= 1000 \end{aligned} \right\} \tag{5.22}$$

3. The technological objectives of a rotor spinning machine are drafting, twisting, and packaging. For each technological function, there is a mechanism, namely, the toothed rafting mechanism, the twisting mechanism, and the building mechanism. Each mechanism has its own power requirements. The summation of these three powers gives the total rotor spinning machine power.

4. When the rotor spinning machine runs with cotton material, its power will be no-load power, but when it operates with cotton slivers to produce yarns, then the machine's power will be full-load power.

   The difference between full-load power and no-load power is the material processing power that is required to transfer a second drawn sliver of virgin cotton to a rotor-spun yarn, wound on a package (spool type).

5. The rotor's productivity is in kilograms per hour. P_0 is calculated by the formula:

$$P_0 = \frac{0.06 \times \pi_R}{\alpha_{tex}} \times \zeta(100\%) \tag{5.23}$$

Active example 8

Calculate the rotor's productivity and its SE when spinning 25 tex yarn with $\alpha_{tex} = 450$, rotor speed = 45 kRPM, rotor $\varphi = 40$ mm, and wharves of rotor-spindle $\varphi = 18$ mm.

Solution

The productivity P_0 is calculated by Formula 5.23:

$$P_0 = \frac{0.06 \times 45,000 \times 25^{3/2}}{4,500} \times \frac{10^{-3} kg}{h}(\zeta = 100\%)$$

$$= 0.075 \text{ kg/h}$$

From Equation 4.15, the rotor's power is

$$P = 36 \times 0.04 \times 0.018 \times 45^2$$

$$= 52 \text{ W}$$

Then, the SE is

$$SE = \frac{0.052}{0.075}$$

$$= 0.693 \text{ kWh/kg}$$

Then, we calculated SE without returning to Equation 4.10.

6. The power distribution of a rotor spinning machine can be graphed for the three mechanisms of the machine using a pie chart or a Sankey diagram.

7. The two main powers required in the rotor spinning machine are the rotor and the air circulated in and around the rotor via the machine fan.

8. The relationship between yarn counts and their twists can be expressed as

$$\text{tpm} = \alpha_m \times \sqrt{N_m}$$

$$\left. \begin{aligned} \alpha_{\text{tex}} &= \frac{\text{tpm}}{\sqrt{\text{tex}}} \\[1em] \text{tpi} &= \alpha_{ec}\sqrt{\text{Nec}} \\[1em] &= \alpha_e\sqrt{\text{Ne}} \end{aligned} \right\} \qquad (5.24)$$

$$\text{tpm} = 39.37 \times \text{tpi}$$

where:

tpm and tpi	= Turns per meter and turns per inch, respectively
N_m, tex, and Ne	= Yarn cotton count in metric, Tex, and English systems
α_m, α_{ec} (α_e), and α_{tex}	= twist factors in metric, Tex, and English systems

Review Questions

Q#1: What are the main mechanisms of the rotor spinning machine? Which of these mechanisms require a large value of watts?

Q#2: What is meant by rotor-spun yarn's specific energy (SE)?

Q#3: Differentiate between the power requirements of both yarn withdrawal from the rotor and yarn withdrawal from the doffing tube via delivery rolls.

Q#4: Answer true or false for the following statements:

a. The rotor-spun yarn twist factor is greater than the twist factor of the same ring-spun yarn.

b. The rotor's power requirements are of the largest value.

c. The no-load power of the rotor spinning machine is smaller than the full-load power of the same machine.

d. Yarn cotton counts ≤10 require less power when processed by a rotor machine compared to a ring spinning machine.

e. The air drag power and air fiber transport power are relatively small as calculated via fan power per rotor.

f. The combing roller speed is higher than the rotor speed.

g. The combing roller speed is higher for artificial sliver toothed drafting than for medium virgin cotton slivers or bony staple cotton.

h. The rotor or ring-spun yarns' SE can be calculated without returning to Formulae 5.18 and 5.19.

i. The SE of rotor-spun yarn of cotton count >30 is lower than the SE of the same ring-spun yarn.

Answers to Review Questions

Q#1: See Summary Point #3. The greatest value power requirements are returned to the twisting mechanism, which requires about 60%–80% of the spinning mill power.

Q#2: See Section 5.9.

Q#3: For yarn's effects on the rotor, see Section 5.3.1; for yarn's withdrawal from the doffing tube, see Section 5.4.

Q#4: (a) true, (b) true, (c) true, (d) false, (e) false, (f) false, (g) true, (h) false.

Review Problem and Solution

If you are given the following data concerning a rotor spinning box (spinning position), calculate the power required for each movable part.

- Yarn tex = 20
- Sliver htex = 5
- α_m = 153.7
- Rotor RPM = 50,000
- Rotor φ = 40 mm

Assume any of the required data.

Solution

A. For opening mechanism power:

$$P = 50 \times d'_\omega \times' n_0$$

$$= 50 \times 0.024 \times 6 \text{ (wharves } \varphi = 24 \text{ mm)}$$

$$= 7.2\,\text{W}$$

B. For the twisting mechanism, the rotor's power is

$$P = 36 \times d'_R \times d'_\omega \times' n_R^2$$

$$= 36 \times 0.04 \times 0.018 \times 50^2 \text{(wharves } \varphi = 18\,\text{mm)}$$

$$= 65\,\text{W}$$

C. For the winding mechanism (yarn withdrawal from doffing tube + package winding):

$$\therefore P_\omega = 0.29 \times d_R^2 \times A_R^2 \times \text{tex} \times V_L \times 10^{-4} + 0.02 \times V_L \times \frac{h'}{d'}$$

$$= 0.29 \times 0.040^2 \times 50^2 \times 20 \times V_L \times 10^{-4} + 0.02 \times V_L \times \frac{300}{200}$$

(package length = 300 mm and package φ = 200 mm)

$$= V_L[0.00232 + 0.02]$$

$$= V_{Lx}$$

$$\text{tpm} = \alpha_m \sqrt{N_m}$$

$$\text{tex20} = N_m 50$$

$$\therefore \text{tpm} = 153.7 \times \sqrt{50}$$

$$= 1087$$

$$\therefore 1078 = \frac{n_R}{V_L}$$

$$\therefore V_L = 46 \text{ m/s}$$

$\therefore$by substituting in P_ω, we get

$$\therefore P_\omega = 46 \times 0.0223$$

$$= 1.026 \text{ W}$$

Then, the total no-load power for the rotor spinning units is

$$P_{\text{no-load}} = 72 + 65 + 1.026$$

$$= 73.22 \text{ W}$$

The percentage distribution:

$$\text{Opening mechanism} = \frac{7.2}{73.22} \times 100$$

$$= 10\%$$

$$\text{Twisting mechanism} = \frac{65}{73.22} \times 100$$

$$= 89.0\%$$

$$\text{Winding mechanism} = 100 - (89 + 10 + \text{winding})$$

$$= 1\%$$

The construction of a pie chart (see Figure 5.1) requires the following steps:

a. Draw a full circle.

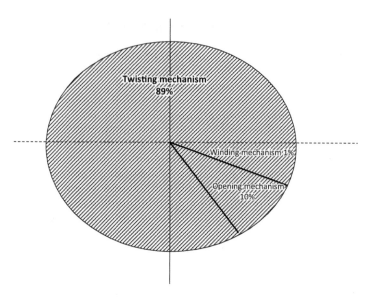

FIGURE 5.1

Π chart for no-load power of a rotor spinning unit. (From Elhawary I.A., 2013, Mechanics of the rotor spinning machine, Lecture notes, TED, Alexandria University, Alexandria, Egypt.)

b. Divide the full central angle 360° by the ratios of the power distribution; the rotor's power will be 360 × 0.89 = 320.4°; the opening mechanism's power will be 360 × 0.1 = 36°; and the winding mechanism's power will be 360 − (320.4 + 36) = 3.6° (see Figure 5.1).

The previous pie chart can be repeated for full-load power.

Bibliography

Moscow State University of Design & Technology (msta.ac), 1975, Rotor's machines, Notes Lab, Moscow, RFU.

Sentil Kumar R., 2015, *Process Management in Spinning*, CRC Press, Taylor & Francis Group, New York.

Reference

Elhawary I.A., 2013, Mechanics of the rotor spinning machine, Lecture notes, TED, Alexandria University, Alexandria, Egypt.

Krause H.W. and Soliman H.A., 1980, The power distribution in a rotor spinning machine, Private communication, Swiss Federal Institute of Technology (SFIT), Switzerland.

6

Air Flow Inside the Rotor Spin Box

6.1 Introduction

It is well known that the air volume that is used to transfer fibers from a combing roller (opening cylinder) to a rotor's sliding wall via a transport duct (tube) must be minimized as much as possible.

To fulfill such a condition, in the case of using a self-pumping rotor, the following parameters must be controlled:

- Rotational speed of the rotor
- Rotor's outside diameter
- Rotor's inside diameter
- Engineered geometrical profile of rotor's circumferential holes
- Engineered geometrical profile of both the fiber transport tube (duct) and the yarn tube (doffing tube)

The first three variables (parameters) control the pressure (vacuum), that is, the difference in the heads of the rotor's circumferential holes. The last two parameters (variables) determine the rate of air flow inside the rotor spin box (unit) (Elhawary, 2014).

6.2 Suction Head of the Self-Pumping Rotor

Shimzu and Takechi (Shirley Institute, 1968) reported that the suction head of the self-pumping rotor can be calculated by applying one of the following formulae:

$$h = 1.9 \times 10^{-8} \times N^2 \left(r_2^2 - r_1^2 \right) \tag{6.1}$$

$$h = 1.9 \times 10^{-8} \times N^2 \left[(2r_2\ell) - \ell^2 \right] \tag{6.2}$$

where:
  h = Suction head in inches of water
  r_1 = Internal rotor's radius in inches
  r_2 = External rotor's radius in inches
  ℓ = Rotor's circumferential hole length in inches
  N = Rotor's revolutions per minute (RPM)

It was found experimentally that when the rotor's circumferential hole length is less than 0.6 in., the theoretical suction head is the same as the practical value of its length. Thus, in the case of a thin-walled rotor, Formula 6.2 will be

$$h = 1.9 \times 10^{-8} \times N^2 \left[1.2r_2 - 0.36 \right] \tag{6.3}$$

This formula is valid when

$$\ell = (r_2 - r_1) \leq 0.6 \text{ in.} \tag{6.4}$$

6.3 Air Flow Rate

It is well known experimentally that the minimum air volume required in a spin box is 1 ft³/min (28.3 L/min), that is, when the rate of air flow is known, then the parameters of the self-pumping rotor can be selected. The relationship between the suction head of the self-pumping rotor and the suction heads for both the fiber transport tube and the yarn tube is

$$h = h_1 + h_2 \tag{6.5}$$

where:
  h = Self-pumping rotor suction head
  h_1 = Suction head loss for both the fiber transport tube and the yarn tube
  h_2 = Suction head loss in the circumferential hole of the self-pumping rotor

6.3.1 Head Loss in Both Fiber and Yarn Tube

Both the fiber transport tube and the yarn tube have head loss in their air flow. This head loss (negative suction) is due to (a) stripping the fibers from the teeth of the toothed combing roll to transport them through a tube to be fed to the sliding wall (inclined wall) of the rotor; and (b) the open-end suction of the yarn inside the rotor, where it forms a yarn arm length that is equal to the rotor's internal radius (doffing tube's navel to the rotor's collecting surface). The negative pressure for this type of head loss for both fiber and yarn tubes is calculated by the formula

$$H = 2\times10^{-2} \times \frac{\mu Q^2}{d_1^5} \times \ell_1 \tag{6.6}$$

where:
  Q = Rate of air flow in cubic feet/min
  d_1 = Tube diameter in inches
  ℓ_1 = Length of tube in inches
  μ = Air drag coefficient (air resistance coefficient), which can be determined by the formula:

$$\mu = 0.0018 + 0.1562\,R_e^{-0.35} \tag{6.7}$$

where R_e is the Reynolds number, which must be less than 4×10^3, that is, its critical value.

$$R_e \leq 4\times10^5 \tag{6.8}$$

The Reynolds number is calculated by

$$R_e = 1.6\times10^3 \times \frac{Q}{d_1} \tag{6.9}$$

where:
  Q = Rate of air low in cubic feet/min
  d_1 = Tube diameter (either fiber or yarn)
  R_e = Reynolds number

6.3.2 Head Loss in the Rotor's Hole

This can be defined as the rotor's suction head (negative pressure), and it means the head loss that is required to enable air flow within the holes of the rotor. It can be calculated by the equation:

$$h_2 = 2.1\times10^{-3} \times \frac{Q^2}{h^2 \times C^2 \times d_2^4} \tag{6.10}$$

where:
  h_2 = Head loss in the rotors circumferential holes
  Q = Rate of air flow in cubic feet/min
  C = Discharge coefficient $=0.6$
  d_2 = Hole diameter in inches

In Formula 6.10, it is assumed that the hole's length is zero, that is, we are dealing with an orifice.

Theoretically, this is required to apply Equations 6.3, 6.5, 6.6, and 6.10.

Active example

For a rotor spinning machine, it was found that for a doffing tube of length 125 mm (4.92") and diameter 4 mm (0.157"), the rotor speed would need $N=35$ kRPM, maximum radius = 28 mm (1.10"), internal radius = 27 mm (1.06"), number of holes = 8, and hole diameter =1 mm (0.04").

Problem:

Calculate the rotor's power due to air flow.

Solution

In this example, we will assume that both pipes of fibers and yarn are of the same diameter (4 mm = 0.157") and of the same length (l = 125 mm/4.42").

Also, we will suppose that Q is the minimum air flow required $Q = 1\,\text{ft}^3/\text{min}\ 4.72 \times 10^{-4}\ \text{m}^3/\text{s}$. Then, the Reynolds number $R_e = 1.6 \times 10^3 \times \dfrac{1}{0.157} = 0.1091 \times 10^5 < 4 \times 10^5$, that is, it is reasonable to apply.

The air drag coefficient $\mu = 0.0018 + \left(0.1091 \times 10^5\right)^{-0.35} = 0.04134$.

The head loss (suction head) of a self-pumping rotor due to its rotational speed 35 kRPM in h (Equation 6.2) is

$$9 \times 10^{-8} \times (35,000)^2 \times \left[1.2 \times 1.02 - 0.6^2\right]$$

$$= 20\,\text{in. of water} = 20 \times 250$$

$$= 5,000\,\text{Pa}$$

The head loss due to air flow inside both the fiber and yarn tubes h (Equation 6.6) is

$$h_1 = 1.2 \times 10^{-2} \times \frac{0.04134 \times 1^2 \times 4.92}{0.157^5}$$

$$= 26\,\text{in. of water} = 26 \times 250$$

$$= 6500\,\text{Pa (for both tubes)}$$

The head loss through the holes of the rotor h_2 (Equation 6.10) is

$$h_2 = 2.1 \times 10^{-3} \times \frac{1^2}{8^2 \times 0.6^2 \times 0.157^4}$$

$$= 0.15 \text{ in. of water} = 0.15 \times 250$$

$$= 37.5 \, \text{Pa}$$

$\therefore h_1 + h_2 = 6538$ Pa. Compare this value to the value of $h = 4968$ Pa.
$\therefore$ If $h_1 + h_2 > h$, then we must consider the largest figure to be 6538 Pa, which means the rotor's additional (extra) power due to air flow is

$$P = Q \times h$$

$$= 4.75 \times 10^{-4} \times 6.538$$

$$= 3.0 \, \text{W}$$

The total rotor's power is

$$P_{\text{total}} = 36 \times 'd_R \times 'd\omega \times 'n_R^2 + Q \times h$$

$$= 36 \times 0.056 \times 0.018 \times 35^2 + 3.0$$

$$= 44 + 3.0$$

$$= 47 \text{ W (94\% [air drag + bearing friction] + 6\% self-pumping)}$$

6.4 Summary Points

1. The self-pumping rotor power P is

$$P = Q.h$$

where:
P = Self-pumping rotor's power in watts
Q = Rate of air flow in m³/s (cubic meter per second)
h = Head loss in pascal, 1 Pa = 1 N/m² (1.0 N·m⁻²)

2. One inch of water (pressure) = 250 Pa.
3. Air flow Q as minimum value is $Q_{\min} = 1 \, \text{ft}^3/\text{min} = 28.3 \, \text{L/min}$ (L/min = 0.472 L/s).

$$= 0.472 \, \text{L/s}$$

$$= 4.72 \times 10^{-4} \, \text{m}^3/\text{s}$$

4. The self-pumping rotor's total power is composed of two parts: part one due to air flow inside the rotor depends on h or $(h_1 + h_2)$, whichever is the biggest figure, and part two is the air friction around the rotor and bearing friction of the rotor spindle as in Equation 4.15.

5. The rotational speed of the self-pumping rotor generates a suction-negative pressure inside the rotor, that is, h due to escape through holes that will lead to air flow from outside the rotor to inside the rotor via the fiber tube and yarn tube (h_1 and h_2). The suction head from the rotor rotational speed in Formula 6.3 is due to air escaping from the rotor's holes to outside, and the combing air flow via the fiber and yarn tube causes head losses $[h_1 + h_2]$, that is, Formulae 6.6 and 6.10.

6. The total head loss is taken from h or $h_1 + h_2$, whichever is the greatest (Table 6.1).

Review Questions

Q#1: Write the required formulae to calculate the power of the self-pumping rotor due to air flow inside the rotor.

Q#2: What is the value of the minimum air flow inside the rotor?

Q#3: What is the critical Reynolds number value for air flow inside the rotor? Write a formula to calculate it.

Q#4: Write a definition for the total power of the self-pumping rotor.

Q#5: Write an equation to calculate the air drag coefficient in the rotor's orifices.

TABLE 6.1

Some Air Properties

Temperature (°C)	Density ρ (kg/m⁻³)	Dynamic viscosity μ (kg m/s)	Kinematic viscosity γ (m²·s¹)
20	1.20	18.15	15.13
40	1.12	19.05	17.01
60	1.06	19.82	18.70
80	0.99	20.65	20.86
100	0.94	21.85	23.24

Source: Elshourbagy, K.A., 2008, Fluid mechanics and hydraulic machines. Lecture, Department of Mechanical Engineering, Alexandria University, Egypt.

Answers to Review Questions

Q#1: See formula for h (Equation 6.1), h_1 (Equation 6.6), and h_2 (Equation 6.10); we can apply either h or $h_1 + h_2$, whichever is the biggest, to calculate the self-pumping rotor's power due to air flow inside it.

Q#2: The minimum value of Q is 1 ft³/min (4.72×10⁻⁴ m³/s).

Q#3: The critical Reynolds number value is

$$R_{ecr} < 4 \times 10^3$$

The formula to calculate it is in Equation 6.9.

Q#4: The total power of the self-pumping rotor is composed of two parts: part one is due to the bearing friction of the rotor spindle and air drag around the rotor, and part two is due to air flow inside the rotor as shown previously in h, h_1, and h_2.

Q#5: See Formula 6.7.

Review Problem

A rotor spinning machine has a rotor with a maximum diameter of 46 mm (thin-walled rotor) and an RPM = 50,000. Both the fiber tubes and yarn tubes have diameters of 5 mm and length of 130 mm. These measurements are required to calculate

a. The suction head of the self-pumping rotor
b. The head loss in both fiber and yarn tubes
c. The head loss (suction head) in the rotor's holes (8 holes with φ2 mm)
d. Compare (a), (b), and (c). Which is bigger?

Answers to Review Problem

a. By applying Formula 6.2 for a thin-walled rotor:

$$\therefore h = 1.9 \times 10^{-8} \times (50,000)^2 \times [0.9055 \times 1.2 - 0.36]$$

$$= 35 \text{ in. of water}$$

$$= 8750 \text{ Pa}$$

where r_1 and r_2 are substituted in inches.

b. For the value of h_1, apply Formula 6.6; for the Reynolds number, apply Formula 6.9:

$$R_e = 1.6 \times 10^3 \times \frac{Q}{d_1}$$

$$\because Q = 1\,\text{ft}^3/\text{min}$$

$$d_1 = 0.1969''$$

$$\therefore R_e = 0.0813 \times 10^5 < 4 \times 10^5$$

As the Reynolds number is correct, we can proceed.
For the coefficient of air drag μ, apply Formula 6.7:

$$\therefore \mu = 0.0018 + 0.1562 \times R_e^{-0.35}$$

$$= 0.04280$$

$$\therefore h_1 = 1.2 \times 10^{-2} \times \frac{0.04280 \times 1^2}{(0.1969)^5} \times 5.118$$

$$= 9\,\text{in. of water}$$

$$= 2.250\,\text{Pa}$$

c. For the h_2 value, apply Formula 6.10:

$$h_2 = 2.1 \times 10^{-3} \times \frac{1}{8^2 \times 0.6^2 \times 0.07874}$$

$$= 2.0\,\text{in. of water} = 500\,\text{Pa}$$

For comparison:
   Therefore, for calculating the power, we must consider 8.160 Pa. The self-pumping rotor's power due to air flow inside is P:

$$P = Q \times h = 4.72 \times 10^{-4} \times 8750$$

$$= 4.0\,\text{W}$$

The rotor's power due to bearing friction and air friction is P:

$$P = 36 d'_R \times d'_w \times n_R^2$$

$$= 36 \times 0.046 \times 0.018 \times 50^2$$

$$= 75\,\text{W}$$

The total power of the self-pumping rotor P_{total} is

$$P_{total} = P_t = 75 + 4 = 79.0 \, W$$

The percentage distribution is 5% (air flow inside the rotor) and 95% (bearing friction and air drag).

Bibliography

Elhawary, I.A., 2015, Rotor's dynamics, Lecture notes, TED, Alexandria University, Alexandria, Egypt.

References

Elhawary I.A., 2014, Dynamic balancing of textile rotating masses, Post-graduate course, TED, Alexandria University, Alexandria, Egypt.
Elshourbagy K.A., 2008, Fluid mechanics and hydraulic machines, Lecture, Department of Mechanical Engineering, Alexandria University, Alexandria, Egypt.
Shirley Institute, 1968, Break spinning: The final report of a three years investigation undertaken by the Cotton, Silk and Man-Made Fibres Research Association, Manchester, UK.

7

Rotor-Spun Yarn's Dynamic Tension

7.1 Introduction

In rotor spinning, the lower end of the rotor-spun yarn is always directed toward the collecting surface of the rotor to be spliced and then combined with the fibers that are deposited in the rotor grooves. The lower part of the rotor-spun yarn from the doffing tube navel (hub) to the rotor groove is called the *yarn arm*. This arm rotates with a rotational speed that is a little bit higher than the rotor rotational speed; therefore, it is subjected to a centrifugal force that will reflect the outgoing tension of the rotor yarn, especially at the outlet of the yarn tube or doffing tube (maximum value). The yarn arm profile is curved to form a rotor balloon; usually, the rotor balloon is curved toward the opposite direction of yarn arm rotation or rotor rotation, that is, the maximum balloon radius is in opposition to the rotor's radian speed ω_R.

7.2 Theoretical Dynamic Yarn Tension Calculations

In this subtopic, we will apply only two of many techniques to calculate the theoretical dynamic tension of yarn.

7.2.1 First Technique

In the first technique, we will assume that the yarn arm is a straight line (Figure 7.1) with length $= D_R/2$ (rotors radius) and a rotational speed with a radian of ω_R. Then, the rotor collecting surface (rotor groove) speed that is equal to the linear speed of point P (extreme end of the yarn arm) can be called V_P:

$$V_P = 60 \times \omega_R \times [\sigma_R/2]$$

$$= \pi.'D_R.N_R \ m/min \tag{7.1}$$

$$= 0.0523 \ 'D_R \times N_R$$

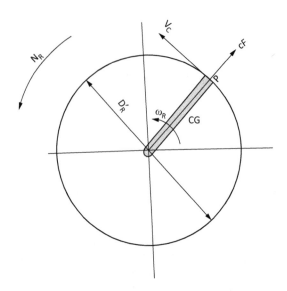

FIGURE 7.1
Straight profile of yarn's arm inside the rotor. CG, yarn arm's center of gravity (mass center); ω_R is the rotor's radian speed in seconds (s); D_R' is the rotor diameter; cF is the yarn arm's centrifugal force (inertia force); V_c is the rotor's groove surface speed; N_R is the rotor's revolutions per minute (RPM). (From Soliman H.A., 1971, Lectures notes, TED, Alexandria University, Alexandria, Egypt.)

where:

> V_P = Linear speed of release point P, which is the circumferential speed of the rotor's groove
> D_R = Rotor's diameter in meters divided by 1000

The yarn arm has its mass center at its half-length and lies at $D_R/4$, at the effective point of the yarn arm's centrifugal force. The direction of the force is outward and coincides with the arm axis and its value cF is

$$cF = \frac{w}{g}\left(\omega_R^2 \cdot D_R /4\right) \tag{7.2}$$

where w is the yarn arm weight that is calculated by

$$w = \left[\text{yarn tex } 10^{-3}\right] \times \frac{D_R}{2} \text{ cN} \tag{7.3}$$

where:

> w = Yarn arm weight in cN
> D_R = Rotor diameter in meters
> g = Acceleration gravity

From Equations 7.2 and 7.3, we can write

$$cF = \frac{tex}{g} \cdot \omega_R^2 \cdot \frac{'D_R}{8} \times 10^{-3} \text{ cN} \qquad (7.4)$$

where cF = yarn arm centrifugal force = arm tension

$$\therefore cF = \frac{tex}{g} \times 10^{-3} \times \left[\omega_R^2 \times 'D_R^2/4\right] \times \frac{1}{2}$$

$$= \frac{tex}{2g} \times 10^{-3} \times V_P^2 \text{ cN} \qquad (7.5)$$

where:

tex	=	Rotor-spun yarn tex
V_P^2	=	Linear speed of release point P

Formula 7.4 can be rewritten as follows after substituting cF with T (yarn tension):

$$T = \frac{tex}{2g} \times 10^{-3} \times V_P^2 \text{ cN}$$

$$\therefore 'T = \frac{T}{tex} \qquad (7.6)$$

$$= \frac{1}{2g} \times 10^{-3} \times V_P^2 \text{ cN/tex}$$

where:

'T	=	Yarn tension in cN/tex
g	=	Gravitational acceleration $\cong 10 \text{ m/s}^{-2}$

For a rotor spinning machine, if V_P is considered 100 m/s, then Equation 7.6 will be

$$T = 0.5 \times tex \text{ cN} \qquad (7.7)$$

If we suppose that the yarn tex = 30, then $\grave{T}$ is 15 cN; the value of $\grave{T}$ is calculated as follows:

$$\grave{T} = \frac{100^2}{2 \times 10} \times 10^{-3} \text{ cN/tex}$$

$$= 0.50 \left[g = 10 \text{ m.s}^{-2} \right]$$

$$\therefore T = \grave{T} \times \text{tex}$$

$$= 0.5 \times \text{tex}$$

Equation 7.6 can be rewritten as follows:

$$\grave{T} = \frac{T}{\text{tex}}$$

(7.8)

$$= 139 \times 10^{-9} \times \grave{D}_R^2 \times N_R^2 \text{ cN/tex}$$

Formula 7.8 shows the applied tension on the yarn arm inside the rotor as a ratio of the yarn tex. To calculate the dynamic tension at the outlet of the doffing tube, Equation 7.8 must be multiplied by e^{μ} where μ is the coefficient of friction between the rotor-spun yarn and the doffing tube's lower end (navel) and is the angle of contact between the lower yarn and the doffing tube navel, which will be summed as $(\pi/2)$. Then, $e^{\mu\theta}$ will be $e^{\mu\theta/2}$ for cotton yarn.

If $\mu = 0.17$, then $e^{0.17 \times \pi/12} = 1.30$. Thus, Equation 7.8 will be

$$'T = \frac{T}{\text{tex}}$$

$$= 178 \times 10^{-9} \times N_R^2 \times \grave{D}_R^2 \ \frac{\text{cN}}{\text{tex}}$$

$$= 178 \times \text{nano} \times N_R^2 \times \grave{D}_R^2 \ \frac{\text{cN}}{\text{tex}}$$

(7.8a)

$$\cong 180 \times \text{nano} \times N_R^2 \times \grave{D}_R^2 \ \frac{\text{cN}}{\text{tex}}$$

(7.9)

where:
  $\grave{D}_R$ = Rotor diameter divided by 1000
  N_R = RPM of rotor
nano = Constant = 10^{-9}

Equation 7.9 is valid for cotton yarns only. To find the dynamic yarn tension for other types of yarn (wool and artificial fibers), the coefficient of friction value μ must be changed for the processed fibers. For cotton yarns, to find the value of the absolute dynamic tension at the outlet of the doffing tube, Equation 7.8a is multiplied by yarn tex. Then, the absolute tension for cotton yarns is

$$T = 178 \times \text{yarn tex} \times \text{nano} \times N_R^2 \times' D_R^2 \, \frac{cN}{tex} \qquad (7.10)$$

where:
  T = Absolute yarn dynamic tension in cN
  D_R = Rotor diameter in meters or divided by 1000
  N_R = Rotor RPM

Active example 1
  A rotor spinning machine with a rotor φ46 mm and an RPM of 45,000 is producing a 40 s count cotton yarn. What will be the dynamic yarn tension at the outlet of the yarn tube?

Solution
  If yarn tex = 24, then T is

$$T = 1.78 \times 24 \times 10^{-9} \times (45,000)^2 \times 0.046^2$$

$$= 180 \, cN$$

In Equation 7.10, if we insert 180 instead of 178, we get T = 18.3 cN, which has a percentage error of 1. Therefore, we can finally apply Equation 7.10:

$$T = 180 \times \text{tex} \times \text{nano} \times N_R^2 \times' D_R^2 \qquad (7.11)$$

The items of the formula have been defined previously.

Active example 2
  Repeat Active Example 1 using a rotor φ33 mm, an RPM of 130,000, and a yarn count of 40.

Solution
  By applying Formula 7.11:

$$T = 180 \times 15 \times 10^{-9} \times (130.000)^2 \times (0.033)^2$$

$$= 49.7 \cong 50 \, cN$$

To calculate yarn tension as a ratio of its tex:

$$\grave{T} = \frac{T}{tex} = 3.3 \, cN/tex$$

In rotor spinning, it is preferable to have a yarn dynamic tension per tex of 0.5, that is, $T = 0.5$ (Formula 7.9). To satisfy this condition for Active Example 2, then

$$180 \times 10^{-9} \times N_R^2 \times (0.033)^2 = 7.5 (15/2 = tex12)$$

$$\therefore N_R = 50.505 \, RPM$$

$$\cong 51 \, kRPM$$

7.2.2 Second Technique

Figure 7.2 shows that the yarn arm has a curved profile (rotor balloon or rotor-spun yarn balloon); the centrifugal tension or the balloon tension is T_0:

$$T_0 = \frac{m \, \omega_R^2 \times R^2}{2} \, cN$$

$$= 0.5 \times m \times \omega_R^2 \times R^2$$

(7.12)

where:

T_0 = Rotor balloon tension
ω_R = Rotor's radian speed in seconds
R   = Rotor diameter in centimeters
m   = Yarn mass permit length (g/cm.s²/cm)

The yarn's centrifugal tension will be magnified by the value of $e^{\mu\theta}$ at the outlet of the doffing tube.

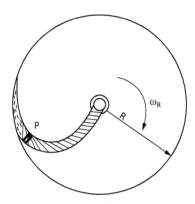

FIGURE 7.2
Curved yarn arm inside the rotor: Rotor balloon. ω_R is the rotor's radian speed in seconds; P is the release point of the fiber from the rotor groove; R is the radius of the rotor. (From Stadler H.W., 1975, *Influence of Rotor Speed on the Yarn Manufacturing Process*, Textile Trade Press, Manchester, UK.)

∴ The dynamic tension T at the outlet of the doffing tube is

$$T = T_0 \times e^{\mu\theta}$$

$$= \left[m.\omega_R^2 \times \frac{R^2}{2} \right] \times e^{\mu\pi/2} \left(\theta = \frac{\pi}{2} \right) \tag{7.13}$$

$$T = 0.5 \times m \times \omega_R^2 \times R^2 \times e^{\mu\pi/2} \ [cN]$$

where:
m = Yarn mass per centimeter $(g/cm.s^{-2}/cm)$
ω_R = Rotor's radian speed in seconds
R = Rotor radius in centimeters
μ = Coefficient of friction between tube navel and yarn = 0.17 for cotton yarns

Formula 7.13 can be altered to take on the following new profile:

$$T = 182 \times 10^{-6} \times \frac{N_R^2 \times D_R^2}{Nm} \ cN \tag{7.14}$$

where:
T = Cotton yarn dynamic tension at the outlet of the doffing tube
N_R = Rotor RPM
D_R = Rotor diameter in meters
Nm = Yarn metric count

To calculate the yarn tension in Rkm or in cN/tex, we multiply yarn tension by Nm = T × Nm (m); for the tension to be in kilometer Rkm, we must divide by 1000. Then, Formula 7.14 will be

$$\grave{T} = 182 \times 10^{-9} \times N_R^2 \times D_R^2 \ cN/tex \tag{7.15}$$

By taking the average values of Formulae 7.15 and 7.8a:

$$T = 180 \times nano \times N_R^2 \times D_R^2 \ Rkm \tag{7.16}$$

If we assume that μ for an artificial fiber (polyester) = 0.50, then Formula 7.16 will be

$$T = 666 \times nano \times N_R^2 \times D_R^2 \ cN/tex \tag{7.17}$$

To get the absolute dynamic tension at the outlet of the doffing tube from Formulae 7.16 and 7.17, we can multiply the product of the formula in yarn-spun tex.

7.2.3 Additional Approaches

a. The relative axial force on the axis of the S_1 spinning rotor is (Rohlena, 1974):

$$\frac{F_0}{tex} = \frac{n_R^2 \times D_R^2}{7.2 \times 10^{12}} \frac{cN}{tex}, RKM, \text{ and } \frac{g}{tex} \qquad (7.18)$$

where:
   F_0/tex = Relative axial forces per tex of yarn (F_0 = Absolute value)
   n_R = Rotor speed in RPM
   D_R = Rotor diameter in millimeters

For example, if $n_R = 30$ kRPM and $D_R = 60$ mm, then $F_0/tex = 0.45$; if the yarn breaking length = 10, then the percentage of the relative axial force on the axis of the rotor is 4.5% of the yarn Rkm. The relationship between rotor speed in RPM and rotor diameter in millimeters is (Rohlena, 1974)

$$n_R = \frac{2.94 \times 10^6}{D_R} RPM$$

For example, if $D_R = 60$ mm, $\therefore n_R = 49$ kRPM.

A proposed equation for the spinning tension is (El Mogahzy and Chewning, 2001)

$$T_{out} = \left[\frac{tex_{yarn} \cdot \hat{\omega}_R^2 \cdot 'D_R^2}{8} \right] . e^{\mu \pi / 2} \qquad \text{(modified)} \qquad (7.19)$$

where:
   T_{out} = Tension of rotor yarn at outlet of doffing tube
   tex_{yarn} = Rotor yarn tex
   $\hat{\omega}_R$ = Radian speed of rotor in seconds divided by 1000
   $'D_R$ = Rotor diameter in millimeters divided by 100
   μ = The coefficient of yarn rotor surface friction
For example, for a cotton yarn with a tex = 60, $n_R = 30$ kRPM, and $D_R = 60$ mm, then

$$T_{out} = 31.74 \text{ cN, i.e., } \frac{cN}{tex} = 0.5293 \cong 0.50$$

7.3 Relationship between $[\sigma/\gamma]$ and V_C

To have a relationship between σ design normal stress in the rotor's material divided by specific weight of rotor's material γ, that is, $[\sigma/\gamma]$ in meters $\left[\frac{N}{m^2} \times \frac{m_3}{N} = m\right]$ and the rotor collective surface (groove) speed V_C in meters per second. This is shown in Figure 7.3.

In Figure 7.3, the horizontal axis represents the circumference speed of the rotor's groove (collecting surface) V_C, where

$$V_C = \pi D_R\, N_R \text{ m/min}$$

$$= 0.052 \times D_R \times N_R \text{ m/s} \qquad (7.20)$$

where:
D_R = Rotor diameter in meters
N_R = Rotor RPM
V_C = Speed in meters per second

The vertical axis on the far left represents σ/γ ; σ is the design stress of the rotor body; and γ is the rotor material's specific weight in N/m³ (N m⁻³). Thus, σ/γ will be in meters.

The second vertical axis on the left represents the yarn tension at the outlet of the doffing tube in cN/tex (gm$_f$/tex) and the right vertical axis represents the rotor RPM.

Active example 3
Refer to Figure 7.3 and use the following data:

$$V_C = 172.8 \text{ m/s}$$

$$\frac{\sigma}{\gamma} = 3000$$

If the yarn dynamic tension at the outlet of the yarn tube = 1.5 Rkm (cN.tex⁻¹−g$_f$.tex⁻¹), find the rotor RPM.

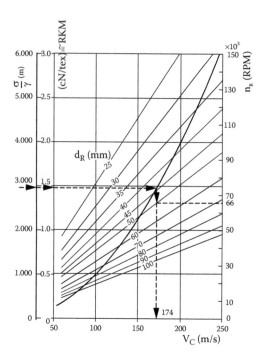

FIGURE 7.3

Rotor groove speed V_C (σ/δ) , rotor RPM, and yarn RKM. V_C is the rotor groove linear speed V_C in meters per second; (σ/γ) is the ratio between rotor design stress σ and its material-specific weight in meters; RKM (N/tex – g/f tex) is the yarn breaking length; n_R is the rotor RPM (CT/min–turns per meter); d_R is the rotor diameter. (Soliman H.A., 1971, Lectures notes, TED, Alexandria University, Alexandria, Egypt.)

Solution

By following or tracking the arrows in Figure 7.3, we find that the rotor's RPM = 66 kRPM (66,000) for a rotor diameter φ50 mm.

7.4 Dynamic Yarn's Tension Measurement

Figure 7.4 shows the concept of a tension meter, which is applied for measuring the outlet yarn dynamic tension of the doffing tube; of course, the value of this tension is not constant as it fluctuates between minimum and maximum values, that is, it has a coefficient of tension variation CV (pa). The yarn tension variability is dependent on the yarn mass variation, that is, count variation (CV) (Uster). The following equation (Stadler, 1975) is determined experimentally:

$$CV(pa) = -0.6 + 0.905\ CV_m\ (Uster) \tag{7.21}$$

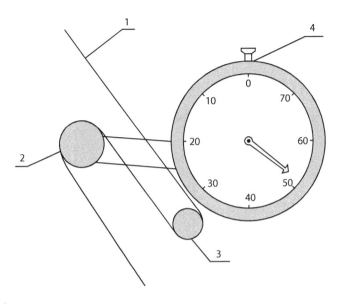

FIGURE 7.4
Measurement concept of rotor-spun yarn's dynamic tension at the outlet of the yarn tube. Yarn at outlet of doffing tube (1); equipment tension rollers (2) and (3); graduated dial in cN (4). (Stadler H.W., 1975, *Influence of Rotor Speed on the Yarn Manufacturing Process*, Textile Trade Press, Manchester, UK.)

where:

CV (pa) = Variability of yarn outlet tension in $CV_m\%$
(pa) = Mean yarn tension
(CV) Uster = Mean variation of yarn mass as measured by Uster tester (UT–5)

7.5 Summary Points

1. The mathematical concept of Figure 7.1 is as follows:

    If the rotor is considered to be a free rotating ring (normal stress σ), then

$$\sigma = \rho\, V_C^2 \left(V_P^2 \right)$$

where:

V_C = V_P = Circumferential speed of rotor's groove or linear speed of release (peeling off point) (Figure 7.1)
ρ = Rotor's material steel ($\rho = 7800\ \text{kg/m}^3$)
σ = Design stress of the rotor in N/m^2 (Pascal [Pa]) from the equation:

$$\sigma = \frac{\gamma}{g} V_C^2$$

where

$$\gamma = \rho \times g$$

where γ is the specific weight of the rotor's material (78,000 Pa):

$$\frac{\sigma}{\gamma} = \frac{V_C^2}{g} \tag{7.23}$$

In Formula 7.6, 'T is

$$T = \frac{1}{2g} \times 10^{-3} \times V_C^2$$

From Equations 7.22 and 7.23, we can find

$$\frac{\sigma}{\gamma} = \frac{2\,T}{tex} \times 10^{-3}$$

$$\frac{\sigma}{\gamma} = \frac{T}{tex} \times 2 \times 10^3 \tag{7.24}$$

With the aid of these different formulae, Figure 7.3 is plotted.

2. The limiting maximum values of the three axes in Figure 7.3 are axis σ / γ is 6 m, RPM axis is 150 RPM, V_C axis is 250 m/s, and rotor's diameter d_R is 100 mm with a minimum rotor diameter d_R of 25 mm.

3. The following formula gives the effect of a rotor's RPM, yarn tex, and yarn twist factor in the tex system on the yarn tension per tex at the outlet of the doffing tube:

$$'T = 1.61 \times 10^{-3} \times 'N_R{}^{2.30} \times tex^{0.33} \times \left('\alpha_{tex}\right)^{0.61} \tag{7.25}$$

where:
$'N_R$ = Rotor's kRPM
$'\alpha_{tex}$ = Yarn twist factor in tex system in thousands
tex = Spun yarn tex

The absolute yarn tension T is

$$T = 'T \times Tex \tag{7.26}$$

4. For Equation 7.21, the fluctuation of yarn tension CV (Pa) ranges from 10% to 17% with the yarn twist factor ranging from 3830 to 6703 (tex system) and the yarn count from 8.5 to 30.5 with a rotor RPM of 40,000.

Active example 4

Using yarn tex = 20, $\alpha_{tex} = 3830$, and rotor RPM = 40,000, apply a formula to calculate $\dot{T}$ and T

Solution

$$\dot{T} = 1.61 \times 10^{-5} \times 40^{2.30} \times 20^{0.33} \times 3.830^{0.61}$$

$$\cong 0.50 \text{ cN/tex}$$

$$\therefore T = 0.50 \times 20$$

$$\cong 10 \text{ cN}$$

5. The yarn arm inside the rotor is the distance from the navel to the collecting surface (rotor groove) and is equal to the rotor radius.

6. If the peeling-off point—release point—rotates at a higher speed than the rotor, this means that it rotates in the same direction as the rotor rotation.

7. The two techniques used to calculate the dynamic yarn tension of the outlet of the doffing tube create the same formula, that is, if it can be assumed that the yarn arm has a straight profile.

8. It is preferable for $\dot{T}$ to be $\cong 0.5$ cN/tex or about 10% of the yarn breaking length (maximum allowable value).

9. The calculation of the rotor-spun yarn's dynamic tension at the outlet of the doffing tube is dependent mainly on the centrifugal tension of the yarn arm inside the rotor—either straight or curved. In both cases, the calculations neglect both the air drag on the yarn arm and the Coriolis force.

Review Questions

Q#1: Define each of the following: yarn arm, yarn peeling-off point, and V_C (V_P).

Q#2: Write the formula to calculate the rotor-spun yarn's dynamic tension at the exit of the yarn tube. Is there any difference in this formula depending on its derivations?

Q#3: Write the formula that considered as a concept of Figure 7.3 chart.

Q#4: Write an equation that describes the value of $\grave{T}$ and T tensions depending on the rotor's speed, yarn tex, and yarn twist factor.

Q#5: Mention an equation that describes the relationship between yarn tension variations in CV (Pa) and yarn mass variation in CV_m (Uster).

Answers to Review Questions

Q#1: See Summary Points 1, 5, and 6 and their corresponding subtopics.

$V_C = V_P =$ Circumferential speed of rotor groove in meters per second, where $V_C = 0.052 \times \grave{D}_R \times N_R$ m/s.

$V_P =$ The linear speed of the peeling-off point $= V_C$.

Q#2: See Equations 7.8a, 7.9, and 7.15. The derivations or approaches are a little bit different, but the final results are the same.

Q#3: See Summary Point 1.

Q#4: See Equation 7.25 in Summary Point 2.

Q#5: See Equation 7.21.

Review Problems

Q#1: A rotor spinning machine is fitted with a rotor with a φ40 mm and an RPM of 80,000 and produces cotton yarn with a tex of 30, calculate $\grave{T}$ and T.

Q#2: Repeat Problem 1 when the yarn twist factor is 3.20 in the English system.

Q#3: By using the $(\sigma/\gamma) - V_C$ chart, calculate the (σ/γ), Rkm, and RPM of a rotor φ45 mm and $V_C = 173$ m/s.

Q#4: Calculate the yarn tension CV (Pa) for yarn number 20.5 when CV_m (Uster) is 18%.

Answers to Review Problems

Q#1: $\grave{T} = 180 \times nano \times N_R^2 \times \grave{D}_R^2$ cN / tex

$= 180 \times 10^{-9} \times (80.000)^2 \times 0.040^2$

$= 1.84$ cN/tex $\left(\text{preferably} = 0.50\right)$

$$\therefore T = 1.84 \times 30$$

$$= 55.3 \text{ cN}$$

Q#2: Use summary points and Formula 7.25.

$$'T = 1.61 \times 10^{-5} \times \grave{N} R^{2.3} \times tex^{0.33} \times \grave{\alpha} \, tex^{0.61}$$

$$= 1.61 \times 10^{-5} \times 80^{2.3} \times 30^{0.33} \times (3.2 \times 957.5)^{0.61}$$

$$= 2.30 \text{ cN/tex}$$

$$T = 2.30 \times 30$$

$$= 72 \text{ cN}$$

It is important to note the following:

$$\alpha_{tex} = \alpha_e \times 957.5 \cong \alpha_e \times 1000$$

Q#3: Returning to Figure 7.3, we can find the following:
For V_C (horizontal axis) $= 173$ m/s, the vertical line will intersect the bolded curve at $\cong 1.5$ g$_f$/tex (cN/tex); a rotor diameter φ50 mm will make the horizontal line intersect the vertical axis of RPM at 76 kRPM. Thus, the answers are

$$\frac{\sigma}{\gamma} \cong 3000, \text{ RKM} \cong 1.5, \text{ and RPM of rotor} = 76 \text{ kRPM}$$

Q#4: Apply Formula 7.21.

$$CV(pa) = -0.6 + 0.905 \, CV_m \, (Uster)$$

$$= \pm 0.6 + 0.905 \times 18$$

$$= 15.69$$

$$\cong 17\%$$

It is important to note the following:

The yarn CV_m (Uster) of 18% is substituted with 18 because CV (pa) is an empirical formula.

Bibliography

Krause H.W. and Soliman H.A., 1970, Textile Milliand, No. 9, Switzerland.
Moscow State University of Design & Technology (msta.ac), 1977, Design & construction of textile machine, Part II, Lecture notes, Moscow, RFU.

References

El Mogahzy Y. and Chewning C.H., Jr., 2001, *Cotton Fiber to Yarn Manufacturing Technology*, Cotton Inc., Cary, NC.
Rohlena V., 1974, *Open End Spinning*, Literature Press, Prague, Czech Republic.
Soliman H.A., 1971, Lectures notes, TED, Alexandria University, Alexandria, Egypt.
Stadler H.W., 1975, *Influence of Rotor Speed on the Yarn Manufacturing Process*, Textile Trade Press, Manchester, UK.

8

Some Mathematics in Rotor Spinning

8.1 Introduction

This chapter presents some basic equations for rotor-spun yarn production on a rotor spinning machine (Section 8.2) and discusses the back-doubling phenomenon inside the rotor of rotor spinning machines (Section 8.3).

8.2 Basic Equations for Rotor-Spun Yarn Production on Rotor Spinning Machines

This section presents partial and total draft twist contractions as well as solved examples and production calculations.

8.2.1 The Draft (Attenuation)

There are different methods for calculating the total draft, such as

$$\text{Total draft} = \frac{\text{Surface speed of take-up roller}}{\text{Surface speed of feed roller}}$$

If we assume that

v_d = Surface speed of delivery roller (take-up roller)

v_f = Surface speed of feed roller (speed of fed sliver to the opening cylinder [toothed cylinder])

then,

$$\therefore D_T = \frac{v_d}{v_f} = \frac{\pi d_d . n_d}{\pi d_f . n_f} = \frac{d_d}{d_f} \times \frac{n_d}{n_f}$$

where:

D_T = Total draft

d_d and d_f = Diameters of delivery and feed rollers, respectively

$\dfrac{n_d}{n_f}$ = Speed ratio between delivery and feed rollers, respectively

This ratio needs to be calculated from the gearing diagram of the rotor spinning machine.

Another method of calculating the total draft via the partial drafts' distribution from the feed roller to the delivery roller will be shown in the following paragraphs.

- The draft between the opening cylinder and the feed roller is the largest value of all the partial drafts. It is called the *mechanical draft D_M*.
- The draft between the fiber tube (duct) and the opening cylinder is called the *stripping draft D_s*.
- The air draft depends on the air velocity at the tube outlet and the air speed at the tube inlet and is symbolized by D_A.

The air speed at the fiber tube inlet is too near in value to the surface speed of the opening cylinder. The value of the air draft inside the fiber tube D_A ranges from 4 to 10 in some machines and from 40 to 45 in other cases. For new, more modern machines, it may have other values.

- The draft between the rotor collecting surface (groove) and the front end of the fiber tube is named the *transport draft D'_T*. It is a little larger than unity. In some cases, we can merge both the air draft inside the fiber tube and the transport draft in a combined figure (value), which is then named D_T^*.
- There is a draft between the lower ends of the doffing tube (yarn tube), that is, the navel and the collection surface of the rotor. The value of such a draft is lower than unity and is therefore called *condensation* and will be defined as follows.

    Condensation is very necessary to form or build up the rotor yarn layers of collecting surface fibers, that is, the successive layers of the fibers during their collection or condensation will generate the whole yarn body at the peeling-off point (critical point between the fibers phase and the yarn phase).

The relationship between the total draft D_T and the partial drafts is

$$D_T = D_M \times D_S \times D_A \times D_T^* \times C \tag{8.1}$$

By applying the *continuity equation* of fluid flow inside a certain path, we can write

$$T_s.V_f = T_0.V_0 \tag{8.2}$$

where:

T_s = Fed sliver tex
V_f = Surface speed of feed roll
T_0 = Tex of fibers layer on the opening cylinder
V_0 = Surface speed of opening cylinder

In Equation 8.2, if the values of T_s, V_f and V_0 are well known, then we can calculate T_0, that is, the tex value of the fiber layer on the surface of the opening cylinder (toothed roller). Also, by knowing the fiber tex of the processed cotton or any other type of fibers, we can calculate the number of fibers in the fibers layer (web) on the surface of the roller, and even the number of fibers inside the sliver can be determined.

In Equation 8.2, the continuity equation can be repeated between two other adjacent elements inside the spin box (spin position), for example, between the doffing tube (yarn tube) and the rotor groove (collecting surface), as follows:

$$V_d.T_y = V_c.T_c \tag{8.3}$$

where:

V_d = Delivery speed or take-up speed
V_c = Surface speed of rotor's collecting surface
T_c = Tex of fibers wedge (fibers strand) on the collecting surface

Equations 8.2 and 8.3 can be written as follows:

$$T_s \times V_f = T_0 \times V_0 = T_c \times V_c = T_y \times V_d \tag{8.4}$$

This is due to the continuity equation of the fluid flow inside the pipes in its general form:

$$Q = A_1 \times V_1 = A_2 V_2 = \text{const} \tag{8.5}$$

where:

Q = Discharge of the fluid from the tube as a rate of flow
A_1, A_2 = Cross-sectional area of two sections of a certain pipe
V_1, V_2 = Fluid speed inside two different sections of a certain tube

For the material flow inside the spin box (spin position), the material tex represents the cross-sectional area of the pipe, and the material flow speed represents the fluid speed inside the tube. In any continuity equation, we can calculate the tex of the material in a certain zone and consequently the number of fibers in a cross section for any zone can be estimated, for example, the number of fibers in the yarn cross section is

$$\frac{T_y}{T_f} \tag{8.6}$$

Finally, the velocity distribution, draft distribution, tex distribution, and distribution of fibers in a cross section can be represented graphically as shown in Figures 8.1 and 8.2.

It is important to note the following:
The total draft can be calculated by

$$D_T = \frac{T_s}{T_y} = \frac{N_{my}}{N_{ms}} = \frac{N_{ey}}{N_{es}} \tag{8.7}$$

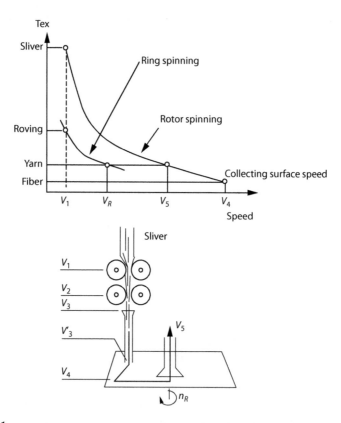

FIGURE 8.1
Continuity equation. V_1–V_2, drafting system 2/2; V_1, back speed (sliver speed); V_2, front speed (=V_R); V_3, inlet air speed; V_3', outlet air speed; P, peeling-off point; V_4, collecting surface speed; V_5, yarn speed. (From Soliman H.A., 1980, Lecture notes, TED, Alexandria University, Alexandria, Egypt.)

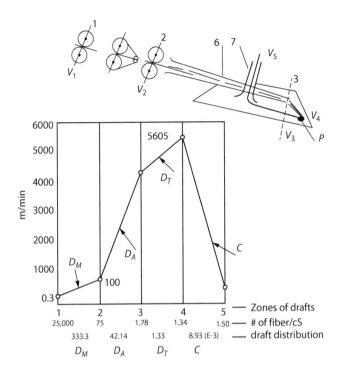

FIGURE 8.2
Draft distribution in a rotor spinning machine. (From Soliman H.A., 1980, Lecture notes TED, Alexandria University, Alexandria, Egypt.)

where:

T_s, T_y = Sliver and yarn texs, respectively
N_{my}, N_{ms} = Metric counts of yarn and sliver, respectively
$\dfrac{N_{ey}}{N_{es}}$ = English counts of yarn and sliver, respectively

Legend for Figure 8.2

$[V_1 - V_2] - \dfrac{3}{3}$ = Drafting system
V_1 = Back roll speed (V_f)
V_2 = Front roll speed
6 = Fiber tube
7 = Yarn tube
P = Peeling-off point
3 = Lower end of fiber tube
V_3 = Air speed at the outlet of fiber tube
V_4 = Rotor collecting surface speed [V_c]
V_5 = Yarn speed (delivery speed) [V_d]

Number of fibers in sliver cross section $= 25{,}000$

Number of fibers in yarn $= 150$

$D_M =$ Mechanical draft $= 333.3$

$D_A =$ Air draft $= 42.14$

$D_T' =$ Transport draft $= 1.33$

$C =$ Condensation $= 8.9 \times 10^{-3}$

$D_T =$ Total draft

$\quad = D_M \times D_A \times D_T' \times c$

$\quad = 167$

$\quad = \dfrac{25{,}000}{150}$

8.2.2 The Twist

The direction of twist S and Z can be determined as follows:

If the direction of the vector of rotor rotation coincides with the direction of the delivery speed (take-up speed) of the rotor-spun yarn, then the twist direction will be S—that is, the vector of ω_R or n_R will be upward.

This is vice versa for the Z direction where the vector of rotor rotation will be opposite to the direction of delivery speed V_d, that is, the vector of ω_R or n_R will be downward.

In all cases, the delivery speed vector will be forward where the rotor yarn is built on the cheese at the top of the rotor machine.

The value of twist is calculated by one of the following methods:

a. First method (approximated)

$$\text{tpm} \cong \frac{n_R}{V_d}$$

where:

tpm $=$ Turns per meter of yarn or twist per meter of yarn

$n_R \quad =$ Rotor revolutions per minute (RPM)

$V_d \quad =$ Delivery speed in meters per minute

b. Second method (accurate)

The peeling-off point (stripping point), that is, the point of inter-action between the yarn's open end and the wedge of fibers on the collecting surface, can be used to measure the twist value (Figure 8.3).

The absolute rotating speed of point P, that is, n_p, is calculated by the formula:

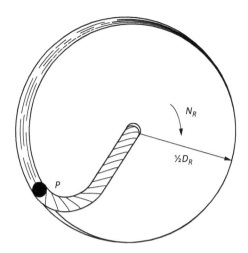

FIGURE 8.3
Peeling-off point P inside the rotor. N_R, rotor's revs per minute; D_R, rotor's diameter; P, peeling-off, stripping, and sweeping point. (From Stadler H.W., 1975, *Influence of Rotor Speed on the Yarn Manufacturing Process*, Textile Trade Press, Manchester, UK)

$$n_p = n_R + \frac{V_d}{nD_R} \tag{8.8}$$

where:
n_p = Peeling-off point's absolute speed in RPM
n_R = Rotor's RPM
V_d = Delivery speed in meters per minute
D_R = Rotor diameter in millimeters

∴ Turns per meter (tpm)

$$\text{tpm} = \frac{n_p}{V_d} \tag{8.9}$$

$$= \frac{\pi R}{V_d} + \frac{1}{\pi D_R}$$

The tpm can be changed to turns per inch (tpi):

$$\text{tpm} = 39.37 \times \text{tpi} \tag{8.10}$$

In Equation 8.9, due to the first approximated method, the difference between the approximated and accurate methods for twist calculation is that the value of $1/\pi D_R$ can be practically neglected.

The following relationships are important:

- tpm

$$\text{tpm} = \alpha_m.\sqrt{N_m} \tag{8.11}$$

where:
α_m = Metric twist factor of the yarn
N_m = Yarn metric count

- tpi

$$\text{tpi} = \alpha_e.\sqrt{N_e} \tag{8.12}$$

where:
α_e = English twist factor of the yarn
N_e = Yarn English count

- tpm (Tex system)

$$\text{tpm} = \frac{\alpha_{\text{tex}}}{\sqrt{\text{Tex}}} \tag{8.13}$$

where:
α_{tex} = Twist factor
tex = Yarn tex

Also,

$$\left.\begin{aligned}
\alpha_m &= 30.3.\alpha_e \\
\alpha_{\text{tex}} &= 31.7.\alpha_m \\
&= 960.5.\alpha_e
\end{aligned}\right\} \tag{8.14}$$

- Shrinkage coefficient η

  The value of the *shrinkage coefficient* is used to compensate for the changes in the yarn tex (due to twist contraction); where the tex increases, usually we modify the increase of its value to decrease the tex so it returns back to the nominal tex. If the nominal delivery speed is V_d, it will be modified to $'V_d$ by the relationship

$$'V_d = \frac{V_d}{\eta}\left('V_d > V_d\right) \tag{8.15}$$

where:
$'V_d$ = Modified delivery speed in meters per minute
V_d = Nominal delivery speed in meters per minute

Active example 1

Calculate the draft distribution, the tex distribution, and the number of fibers per cross section of fiber flux distribution for every moving element in the spin box (spin position of the rotor unit) of rotor spinning machine type BD 200 and take into consideration the following data:

- Yarn tex = 20
- Sliver ktex = 5
- Yarn metric twist factor = 153.7 turns/$(N_m)^{1/2}$
- Rotor RPM = 30,000
- Rotor diameter D_R = 60 mm
- Shrinkage coefficient = 0.95
- Opening cylinder speed = 6 kRPM
- Opening roller (toothed cylinder) diameter D_0 = 65 mm
- Air speed at the inlet of the fiber tube = 23 m/s
- Suppose air speed at the outlet of the fiber duct = 89.7 m/s
- Cotton fiber mtex = 2000

Solution

- *Speed calculations*

As mentioned previously, total draft is

$$D_T = \frac{T_s}{T_y}$$

$$= \frac{5000}{20}$$

$$= 250$$

$$\because \text{tpm} = \frac{153.7 \times 31.7}{\sqrt{20}}$$

$$= 1089$$

$$\therefore V_d = \frac{30.000}{1089}$$

$$= 27.5 \text{ m/s}$$

$$\because D_T = \frac{V_d}{V_f}$$

$$\therefore V_f = \frac{27.5}{250}$$

$$= 0.11 \text{ m/s}$$

The surface speed of the opening cylinder is

$$\pi D_0 n_0 = \pi \times 0.065 \times 6000$$

$$= 1225.0 \text{ m/s}$$

The air speed at the inlet of the fiber tube adjacent to the opening cylinder surface is

$$= 23 \times 60$$

$$= 1380 \text{ m/s}$$

The surface speed of the rotor groove = collecting surface = V_c

$$\therefore V_c = \pi D_R N_R$$

$$= \pi \times 0.06 \times 30{,}000$$

$$= 5{,}654.87 \text{ m/s}$$

$$= 94.25 \text{ m/s}$$

- *Draft distribution (partial drafts):*
- *Mechanical draft D_m*

$$D_m = \frac{V_0}{V_{f0}}$$

$$= \frac{1225.22}{0.11}$$

$$= 11{,}138.36$$

$$\cong 11{,}000$$

- *Stripping draft D_s*

$$D_s = \frac{V_a}{V_0}$$

$$= \frac{1380}{1225.0}$$

$$= 1.1265$$

- Air draft inside the fiber duct (tube)

$$D_A = \frac{\text{Speed of air at outlet}}{\text{Speed of air at inlet}} = \frac{'V_a}{V_a}$$

$$= \frac{89.7 \times 60}{23 \times 60}$$

$$= 3.9$$

$$\cong 4$$

- Transfer draft D_T^*

$$D_T^* = \frac{\pi D_R n_R}{'V_a}$$

$$= \frac{94.25 \times 60}{89.7 \times 60}$$

$$= 1.05$$

If we combine the transfer draft with the air draft, we find

$$D_A \times D_T^* = 3.9 \times 1.05$$

$$= 4.10$$

If we directly divide the surface speed of the rotor groove by the surface speed of the opening cylinder, we get $\frac{94.75 \times 60}{1225.0} = 4.64$ [$\cong$ 10% error] compared to 4.10. Therefore, if we do not have any data about the air speeds in the fiber tube, either at the inlet or outlet, we can combine the stripping draft, air draft, and transport draft in one figure (value).

- Condensation C

$$C = \frac{\text{Surface speed of rotor-spun yarn}}{\text{Delivery speed of rotor collecting surface}}$$

$$= \frac{V_d}{V_c}$$

$$= \frac{V_d}{\pi D_R n_R}$$

$$= \frac{27.5}{5654.87}$$

$$= 4.86 \times 10^{-3}$$

To assume the value of the total draft, then

$$D_T = 11.138 \times 1.126 \times 3.9 \times 1.05 \times 4.86 \times 10^{-3}$$

$$= 249.6$$

$$\cong 250$$

As mentioned previously, if we combine stripping draft D_s, air draft D_A, and transport draft $'D$ as one value, the name will be modified as, $'D_A$, that is,

$$'D_A = D_s \times D_A \times 'D_T$$

This can lead us to consider the total draft inside the rotor spinning machine in three zones:

- Mechanical draft D_M
- Modified air draft $'D_A$
- Condensation C

Thus, the total draft D_T is

$$D_T = D_m \times 'D_A \times C \tag{8.16}$$

It is important to note the following:

- **Shrinkage coefficient**

    If the shrinkage coefficient of the rotor yarn is 0.95 (95%), then the delivery speed V_d will be $'V_d$:

$$'V_d = \frac{V_d}{\eta}$$

$$= \frac{27.5}{0.95}$$

$$= 29.0 \text{ m/s}$$

$$\therefore \text{Total draft} = \frac{29}{0.11}$$

$$= 263.6 \qquad (8.17)$$

Then, the nominal yarn tex $T_y = 20$ will be $'T_y$, where $'T_y = T_y \times \eta$ $= 20 \times 0.95 = 19$. The total draft D_T will be

$$D_T = \frac{5000}{19} = 263.2 \qquad (8.18)$$

By comparing the total draft from speed distribution and material distribution, the percentage error is 5.0 due to the effect of the shrinkage coefficient.

- **The twist calculation**

As mentioned previously, the peeling-off point p, sweeping point p, or stripping point p is responsible for accurate twist insertion. The absolute speed of point p is

$$n_p = n_R + \frac{V_d}{\pi D_R}$$

$$= 30,000 + \frac{27.5}{\pi \times 60 \times 10^{-3}}$$

$$= 30,000 + 146$$

$$= 30,146 \text{ RPM}$$

$$\therefore \text{tpm} = 30,146/27.5$$

$$= 1,096$$

But, if we take into consideration the shrinkage coefficient η, then V_d will be $27.5/0.95 = 29$ m/s.

$$\therefore n_p = 30,000 + \frac{29}{\pi \times 0.06}$$

$$= 30,000 + 154$$

$$= 30,154 \text{ RPM}$$

The tpm (accurate) is

$$\text{tpm} = \frac{30,154}{29}$$

$$= 1,039$$

But the value of tpm differs by 5% due to shrinkage; the η' twist calculation using the approximated method will give a twist difference with a percentage error of 0.50.

- **Tex distribution of fiber groups in different zones (fiber flux)**

If the fed sliver tex T_s is 5000, then the tex of the fibrous web on the surface of the opening cylinder using the continuity equation is

$$T_s \times V_f = T_0 \times V_0$$

(8.19)

$$5000 \times 0.11 = T_0 \times 1225.2$$

∴ $T_0 = 0.4489$ on the surface of the opening cylinder

$$\therefore d_{\text{tex}} = 4.49$$

If we know that the modified air draft is 4.61, then the fiber tex on the collecting surface is

$$T_c \times V_c = T_0 \times V_0 \text{ (continuity equation)}$$

$$\therefore T_c = \frac{T_0 \times V_0}{V_c}$$

$$= \frac{T_0}{D'_A} \left[D'_A = V_c / V_0 \right]$$

$$= 4.49/4.61$$

$$= 0.0974$$

$$\therefore d_{\text{tex}} = 0.974$$

To calculate the yarn tex:

$$T_y \times V_d = V_c \times 0.0974$$

$$\therefore T_y = \frac{94.25 \times 0.0974}{27.5}$$

$$= 20$$

∴Yarn tex T_y is 20

- **Number of fibers per cross section for different fiber strands in different zones of the spin box**

  By assuming that the cotton fiber tex is 0.20, then the number of fibers per cross section in any strand of fibers at any zone is

  $$\frac{\text{Fiber strand tex}}{\text{Fiber tex}}$$

  $\therefore$ Number of fibers in sliver cross section is

  $$= \frac{5,000}{0.2}$$

  $$= 25,000 \text{ fibers per cross section}$$

- The number of fibers per cross section in the fibers' web on the opening cylinder surface is

  $$= \frac{0.4489}{0.2}$$

  $$= 2.2445$$

  $$= 2.25 \text{ fibers}$$

- The number of fibers per cross section in the fiber strand on the collecting surface is

  $$= \frac{0.0974}{0.2}$$

  $$= 0.486 \text{ fibers}$$

- The number of fibers in the yarn cross section is

  $$= \frac{20}{0.2}$$

  $$= 100 \text{ fibers}$$

It is important to note the following:
The fiber strand's tex distribution can be used to calculate the draft distribution as follows:

- The draft between the opening cylinder and feed roll fibers (mechanical draft) D_M is

  $$D_M = \frac{\text{Opening cylinder fibers}}{\text{Feed roller fibers}}$$

$$= \frac{25,000}{0.4489}$$

$$= 11,138$$

- The air draft $'D_A$ is

$$'D_A = \frac{\text{Opening cylinder fibers}}{\text{Collecting surface fibers}}$$

$$= \frac{2.8445}{0.486}$$

$$= 4.61$$

- Condensation C is

$$C = \frac{\text{Collecting surface fibers}}{\text{Yarn fibers}}$$

$$= \frac{0.486}{100}$$

$$= 4.86 \times 10^{-3}$$

- The total draft D_T is

$$D_T = D_M \times 'D_A \times C$$

$$= 11.138 \times 4.61 \times 4.80 \times 10^{-3}$$

$$= 249.54$$

$$= 250$$

c. Production calculation of the rotor spin box

The following equations are used:

- **Tex system**

$$P = \frac{0.06 \times N_R \times \text{tex}^{3/2}}{\alpha_t} \, \text{g/h}$$

where:

P = Rotor productivity in grams per hour (g/h) (theoretical value)

N_R = Rotor RPM.

tex = Rotor-spun yarn tex

α_t = Rotor yarn twist factor in tex system

- **English system**

$$P = \frac{0.0019 \times n_R}{\alpha_e \times Nec^{3/2}} \, \text{lb/h}$$

$$= \frac{0.0019 \times 45,000}{4.5 \times 15^{3/2}}$$

$$= 0.3270 \, \text{lb/h}$$

$$\cong 0.327 \times 453.6 \, (1 \, \text{lb} = 453.6 \, \text{g})$$

$$= 148.0 \, \text{g/h}$$

8.3 Summary Points

1. The basic equations concerning rotor-spun yarn production on rotor spinning machines are

- The total and partial drafts, that is, draft distribution inside the spin box
- The twist calculation
- The speed distribution in meters per minute or meters per second, that is, surface speed
- The fiber strand's distribution as tex distribution
- The number of fibers per cross section of any fiber strands, that is, the distribution of fibers per cross section (fiber flux distribution)
- The productivity of the rotor in grams per hour or pounds per hour
- The draft distribution, tex distribution, and fiber flux distribution can be represented in values or in diagrams (charts, graphs)
- The total draft D_T can be calculated mechanically or technologically:

Mechanical calculation of D_T

$$D_T = \frac{\text{Surface speed of take-up roller (delivery roller)}}{\text{Surface speed of feed roller}}$$

$$= \frac{\pi V_d}{V_f}$$

$$= \frac{\pi n_d . 'd_d}{\pi n_f . d_f}$$

$$= \frac{'d_d}{d_f} . \frac{n_d}{n_f}$$

where:

$'d_d$ and d_f = Diameters of delivery and feed roll, respectively

$\dfrac{n_d}{n_f}$ = Speeds of delivery and feed rolls, respectively

To be calculated, a rotor machine gearing diagram is needed.

Technological calculation of D_T

$$D_T = \frac{T_s}{T_y}$$

$$= \frac{N_y}{N_s}$$

$$= \frac{\text{Number of fibers/cross section of sliver}}{\text{Number of fibers/cross section of yarn}}$$

where:

T_s and T_y = Sliver and yarn texs, respectively

N_s and N_y = Sliver and yarn English counts

- The partial drafts or draft distribution between every two neighboring parts is the ratio between the bigger linear speed and the smaller linear speed, or can be calculated using the tex distribution of fiber flux distribution. Generally, the total draft is the multiplication product of the partial drafts.

- For any surface speed of any rotating elements (e.g., the opening cylinder):

$$V_0 = \pi \grave{D}_0 n_0 \text{ m/min}$$

where:

$\grave{D}_0$ = Opening cylinder diameter in millimeters

n_0 = Opening cylinder RPM

- The absolute peeling-off point P's speed n_p is

$$= n_R + \frac{1}{\pi D_R}$$

where:

n_R = Rotor speed in RPM
D_R = Rotor diameter in millimeters

Usually, it is the stripping point P leading the rotor that is higher in speed. This point is the real twist inserter in the rotor yarn.

- In practice, the rotor yarn's tpm

$$= \frac{n_R}{V_d}$$

where V_d is the delivery speed in meters per minute.

- For rotor productivity P,

$$P = \frac{0.06 \times N_R \times \text{tex}^{3/2}}{\alpha_t} \, \text{g/h}$$

$$= \frac{0.0019 \times N_R}{\alpha_t \times Nec^{3/2}} \, \text{lb/h}$$

where:

P = Rotors' productivity in grams per hour or pounds per hour
N_R = Rotor RPM
Tex and N_e = Rotor yarn tex or English count, respectively

2. Fiber mtex $\quad$ = Fiber tex × 1000
   Fiber dtex $\quad$ = Fiber tex × 10
   dtex (deci tex) = Denier

3. The continuity equation for air flow inside a tube with two cross sections, 1 (big) and 2 (small), is

$$A_1 \times V_1 = A_2 \times V_2 = Q_{\text{discharge}}$$

where:

A_1, A_2 = Air tube cross-sectional areas at the smallest and largest points
V_1, V_2 = Smallest and largest air speeds in cross sections 1 and 2

By analogy, for fiber flow (mass flow) inside the spin box, the tex represents a cross-sectional area of the fiber strands or yarn and speed as in the continuity Formula 8.4, that is, tex versus speed.

4. The air draft inside the fiber tube ranges from 4 to 10 in some machines, and from 40 to 45 in other machines.

5. The air speed at the inlet of the fiber tube is 24 m/s while at the outlet of the fiber tube, toward the sliding wall of the rotor, the air speed is 94 m/s, that is, $D_A = 4$. The fiber transport duct has a conical shape to increase the air speed toward the rotor sliding wall.

6. The modified air draft $'D_A$ is the surface speed of the rotor collecting surface/surface speed of the opening cylinder (toothed cylinder).

Review Questions

Q#1: Write briefly in text format what is meant by the basic equations for rotor-spun yarn production on a rotor spinning machine.

Q#2: Write formulae to calculate the following:

- Mechanical draft
- Modified air draft
- Condensation

What is the value of the total draft?

Q#3: What is the physical meaning of the continuity equation in general, and in the rotor spin box in particular? How can it be applied?

Q#4: Write a formula for calculating the following:

- Both the absolute and relative speeds in RPM of the sweeping point on the rotor groove.
- The tpm using accurate and approximated methods.

Q#5:

Complete the following:

$$tpm = \ldots\ldots\ldots\ldots\sqrt{N_m}$$

$$tpm = \alpha_e\sqrt{\ldots\ldots\ldots\ldots}$$

$$tpm = \frac{\ldots\ldots\ldots\ldots}{\sqrt{tex}}$$

$$\alpha_m = \ldots\ldots\ldots\ldots .\alpha_e$$

$$\alpha_{tex} = \ldots\ldots\ldots\ldots .\alpha_m$$

$$= 960.5 \ldots\ldots\ldots\ldots$$

$$tpm = \ldots\ldots\ldots tpi = \ldots\ldots\ldots$$

Q#6:

- Why is the shrinkage coefficient η important?
- Complete the following:

$$'V_d = \frac{\cdots}{\eta}$$

Q#7:

- What is meant by fiber flux in the rotor spin position?
- Complete the following:

$$\text{Fiber mtex} = \text{Fiber tex} \dots\dots\dots$$

$$\text{Fiber dtex} = \text{Fiber tex} \dots\dots\text{tex}$$

$$\text{dtex} = \dots\dots\dots\dots\dots\dots$$

Q#8: Mention the different equations used to calculate the productivity of a rotor unit in the tex and English systems.

Answers to Review Questions

Q#1: See Section 8.2.

Q#2:

- Mechanical draft D_M is

$$D_M = \frac{\text{Surface speed of opening cylinder}}{\text{Surface speed of feed roller}}$$

$$= \frac{V_0}{V_f}$$

where:
$V_0 = $ Surface speed of opening roller
$V_f = $ Surface speed of feed roller

- Modified air draft D_A^* is

$$D_A^* = \frac{\text{Linear speed of rotor collecting surface}}{\text{Surface speed of opening cylinder}}$$

$$= \frac{V_c}{V_0}$$

where:
V_c = Linear speed of rotor collecting surface
V_0 = Surface speed of opening roller

- Condensation C is

$$C = \frac{V_d}{V_c}$$

where:
V_d = Yarn delivery speed
V_c = Surface speed of collecting surface

The value of total draft D_T is

$$D_T = \frac{V_d}{V_f} = \frac{T_s}{T_y}$$

where:
V_d = Yarn delivery speed
V_f = Surface speed of feed roller
T_s = Sliver tex
T_y = Yarn tex

Q#3: In the air flow inside multiple different cross-sectional areas, the product of $V \times A = Q$ is constant, that is, $V_1 A_1 = V_2 A_2 = V_n a_n = Q$ (discharge). For the mass flow of fibers inside the rotor unit with different cross sections (texs) and speeds, we can write

$$V_1.\text{tex}_1 = V_2.\text{tex}_2 = V_3.\text{tex}_3 \ldots = P(\text{Productivity})$$

Q#4:
- The peeling-off point's absolute speed is

$$n_p + n_R \frac{V_d}{\pi D_R}$$

- Its relative speed is

$$n_p - \pi_R = \frac{V_d}{\pi D_R}$$

$$= {}'n_p$$

where:
n_p = Sweeping point's absolute speed
$'n_p$ = Sweeping point's relative speed
n_R, D_R = Rotor RPM and diameter, respectively

- The accurate tpm is

$$\text{tpm} = \frac{n_p}{V_d}$$

$$= \left(n_R + \frac{V_d}{\pi D_R} \right) \times \frac{1}{V_d}$$

$$= \frac{n_R}{V_d} + \frac{1}{\pi D_R}$$

- The approximated tpm is

$$\text{tpm} = \frac{\pi R}{V_d}$$

where:
V_d = Delivery speed (yarn speed)
n_p = Absolute speed of peeling-off point
n_R = Rotor speed RPM

Q#5:
- $\text{tpm} = \alpha_m \sqrt{N_m}$
- $\text{tpi} = \alpha_e \sqrt{N_{ec}}$
- $\text{tpm} = \dfrac{\alpha_{\text{tex}}}{\sqrt{\text{tex}}}$
- $\alpha_m = 30.3\alpha_e$
- $\alpha_{\text{tex}} = 31.7\,\alpha_m$
- $\phantom{\alpha_{\text{tex}}} = 960.5\alpha_e$
- $\text{tpm} = 39.37\,\text{tpi}$

Q#6: The shrinkage coefficient η is important because we can use it to adjust the nominal delivery speed V_d to $'V_d$ $('V_d = V_d / \eta)$ to increase the draft to have a lower yarn tex, so that when it contracts due to twist, it will be equal to the nominal twist.

Q#7:

- The fiber flux means the distribution of fiber masses (strands) flowing inside the different elements of the rotor spin box (spin position) starting from sliver to yarn.
- Fiber mtex = 1000 tex
- Fiber dtex = 10 tex
- dtex = denier

Q#8: Rotor productivity in the tex system is

$$P = \frac{0.06 \times N_R \times \text{tex}^{3/2}}{\alpha_t} \text{ g/h}$$

$$= \frac{0.0019 \times N_R}{\alpha_e . N_{ec}^{3/2}} \text{ lb/h}$$

where:

P = Theoretical rotor productivity
N_R = Rotor RPM
tex and N_e = Rotor yarn tex and English count, respectively

Review Problems

Q#1: If you are given the following data for a rotor spinning machine, calculate the speed of draft and fiber (flux) distribution:

- Sliver g/m = 4
- Yarn tex = 25
- Opening cylinder = 70 mm
- Opening cylinder speed = 4500 RPM
- Fiber mtex = 200
- Yarn twist factor (tex system) $\alpha_T = 4500$ turns/mtex$^{1/2}$

Represent your answer graphically.

Q#2: Calculate the productivity of the rotor of the rotor spinning machine in Q#1.

We will assume that the air speed at the fiber tube outlet = 6000 m/min (less than 30% of the rotor groove linear speed) (Figure 8.1).

Q#3:

$$P = \frac{0.06 \times 45,000 \times 25^{3/2}}{4,500}$$

$$= 75 \text{ g/h}$$

Answers to Review Problems

Q#1:

- Surface speed of the collecting surface V_c

$$= \frac{45,000 \times 60 \times \pi}{1,000}$$

$$= 4884 \text{ m/min}$$

- Yarn delivery speed V_d

$$= \frac{N_R \times \sqrt{\text{tex}}}{\alpha_t}$$

$$= \frac{45,000 \times \sqrt{25}}{4,500}$$

$$= 50 \text{ m/min}$$

- Surface speed of feed roll V_f

$$= \frac{50 \times 25}{4000} \left[V_f \times T_s = V_d \times T_y \right]$$

$$\therefore V_f = 0.212 \text{ m/min}$$

- Surface speed of opening cylinder V_0

$$= \pi \times 0.07 \times 8000$$

$$= 1760 \text{ m/min}$$

8.4 Back Doubling

The back-doubling phenomenon takes place inside the rotor when the fibers slide down from the rotor sliding wall to the rotor groove to create successive layers of fibers in the shape of a wedge. The maximum number of fibers is at the connection point with the open-end yarn that was previously spun, while the minimum number of fibers is at the other end of the wedge. The minimum is usually zero fibers, that is, the mass flow (fiber flow) breaks behind the peeling-off point (sweeping point). This means that the wedge of fiber is spread on the internal rotor groove's circumference. The building of fiber layers in the wedge generates a new length of rotor yarn. As the number of layers increases, this will lead to more regular yarn, that is, less mass variation ($CV_m\%$). The number of layers means the number of doublings (back doubling). Back doubling is mainly responsible for the improvement of the random drafting waves that are present in the fed sliver and can negatively affect yarn mass variation ($CV_m\%$). Back doubling inside the rotor can also improve the irregularity of the yarn due to harmonic waves in the fed sliver but with certain conditions or restrictions as will be shown later.

8.4.1 Definitions

There are three definitions for back doubling inside a rotor:

- *First definition*

$$\text{Back doubling} = \frac{\text{Linear speed of the rotor groove}}{\text{Yarn delivery speed}}$$

$$= \frac{V_c}{V_d}$$

$$= \frac{\pi D_R . N_R}{V_d} \tag{8.20}$$

$$= \frac{N_R}{V_d} . \pi D_R$$

where:
V_c = Circumferential speed of rotor's collecting surface
N_R = Rotor RPM
D_R = Rotor diameter
V_d = Yarn delivery speed

It is well known that the number of turns (twist) per unit length of yarn T is

$$T = \frac{N_P}{V_d}$$

$$= \frac{1}{V_d}\left[N_R + \frac{V_d}{\pi D_r}\right] \tag{8.21}$$

$$= \frac{n_R}{V_d} + \frac{1}{\pi D_R}$$

where:

    T = Twist per unit length of yarn

    N_p = Absolute velocity of the peeling-off (sweeping) point (Figure 8.3)

  N_R, D_R = Rotor RPM and diameter, respectively

    V_d = Delivery speed of rotor-spun yarn

From Equations 8.20 and 8.21, we can write

$$\text{Back doubling } (BD) = \pi D_R T - 1 \tag{8.22}$$

$\therefore$ BD = Rotor circumference × turns (twist) per unit length − 1

$$\therefore \text{BD} \propto \pi D_R T \tag{8.23}$$

This means that BD equals the number of times the rotor circumference turns per unit length minus one. If the rotor circumference is in meters, we must consider T = tpm. If the circumference is in inches, we consider T = tpi.

- *Second definition*

    As mentioned previously,

$$BD = \frac{V_C}{V_d}$$

$$= \frac{\pi D_R N_R}{\pi D_R .' N_p} \tag{8.24}$$

$$= \frac{\text{RPM}}{\text{Relative RPM of peeling-off point P}}$$

where:

  N_R = Rotor RPM

  $'N_p$ = Relative speed of peeling-off point P's RPM

It has been previously shown that the absolute velocity of peeling-off point P is

$$N_p = N_R + \frac{V_d}{\pi D_R} \tag{8.25}$$

But $'N_p$ is

$$'N_p = \frac{V_d}{\pi D_R} \tag{8.26}$$

$\therefore$ From Equations 8.25 and 8.26,

$$N_p = N_R + 'N_p \tag{8.27}$$

where:

N_p = Absolute RPM of peeling-off point P
$'N_p$ = Relative RPM of peeling-off point P
V_d = Yarn delivery speed
πD_R = Rotor circumference

Equation 8.24 can be rewritten as follows:

$$BD = \frac{N_R}{V_d / \pi D_R}$$
$$= \frac{N_R}{N_p} \tag{8.28}$$

In Equation 8.24, it is clear that back doubling measures the number of revolutions of a rotor for one revolution of peeling-off point P. If we assume that BD = 100, this means that the rotor rotated 100 times for one revolution of the sweeping point P on the rotor circumference.

In other words, back doubling means how many times any point on the rotor circumference has passed in front of the lower end of the fiber tube that is inside the rotor. Again, this will give the number of layers that formed the rotor-spun yarn. If we assume that a rotor yarn is composed of 200 fibers per cross section and the back doubling was 100, then each layer of yarn has two fibers (200/100). This means that only two fibers come out the lower end of the fiber duct (tube), and that the layer of yarn is composed of two fibers.

It has been noted from previous definitions that back doubling is the reciprocal of condensation (C). If the draft from the feed roll to the rotor collecting surface is named spinning draft,

$\therefore$ back doubling $\times$ total draft = spinning draft

$$\therefore BD \times D_T = D_s \tag{8.29}$$

where:

BD = Back doubling inside the rotor (collecting surface)

D_T = Total draft

$$D_s = \text{Spinning draft} = \frac{\text{Surface speed of collecting surface}}{\text{Surface speed of feed roll}} \qquad (8.30)$$

$$= \frac{V_c}{V_f}$$

- *Third definition*

    The third definition will be explained via an active example.

Active example 2

A rotor spinning machine produces a rotor-spun yarn with $N_e = 12$ and tpi = 15. The rotor diameter $D_R = 2.55"$ and the rotor RPM = 30,000. Explain the back doubling.

Solution

The yarn delivery speed (V_d):

$$V_d = \frac{N_R}{\text{tpi}}$$

$$= \frac{30,000}{15} \qquad (8.a)$$

$$= 2,000 \text{ inch/min}$$

If we assume that the rotor is fixed, that is, that there is no rotation while the yarn is withdrawn by V_d, then the peeling-off point will rotate on the rotor collecting surface taking different positions such as P_1, P_2, and so on (Figure 8.4). During the sweeping point P rotation, its RPM will be $'N_p$ (relative speed).

$$'N_p = \frac{V_d}{\pi D_R}$$

$$= \frac{2000}{\pi \times 2.55} \qquad (8.b)$$

$$= 250 \text{ RPM}$$

This means that the rotor rotates (30,000/250) 120 times per revolution of point P. This leads to the concept that any point on the rotor circumference

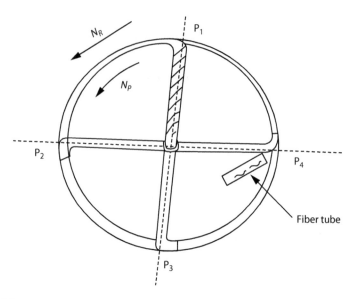

FIGURE 8.4
Motion of point P along the rotor circumference. (From Soliman H.A., 1980, Lecture notes TED, Alexandria University, Alexandria Egypt.)

that passes the front end of the fiber tube toward point P occupies 360°, that is, for each 3° angular movement, the rotor will rotate one revolution.

If we assume that the fiber metric count = 5000 for spun yarn of metric count 20 ($N_e = 12$), then the numbers of fibers per cross section = 250 (5000/20). These fibers are not found once but are built in layers (120) after 120 revolutions of the rotor, that is, the number of fibers per layer = 250/120 = 2.08. This will give the number of fibers that are emerging from the lower end of the fiber tube (2.08 fibers), which will be accumulated in yarn after 120 revolutions of the rotor while point P moves through only one revolution.

It is important to note the following:

- In the previous example, we neglect one additional turn due to the rotation of point P.
- If the rotor circumference $\cong 8''$ and the tpi $= 15$, then $15 \times 8 = 120$, $\left[N_R / {}' N_p = 30{,}000 / 250 \right]$

Thus, BD is

$$\therefore BD = \pi D_R \times tpi \qquad (8.c)$$

$$= Circumferences\ of\ rotor \times yarn\ tpi$$

Figure 8.5 shows every central angle. In the full yarn cross section, 120 layers will be formed.

8.4.2 Back Doubling and Harmonic Waves in the Fed Sliver

Figure 8.6 shows the mass distribution of fed sliver to the rotor machine as a function of the sliver length. It is clear that the distribution follows a cosine curve that has the following formula:

$$m_s = a.\cos\left[\frac{2\pi}{n}.\ell\right] \qquad \text{(Figure 8.1)} \qquad (8.31)$$

where:

$a =$ Amplitude of the cosine curve
$n =$ Wavelength of the cosine curve of the sliver
$\ell =$ General length of the fed sliver
$m_s =$ Sliver mass at any sliver length ℓ

Equation 8.31 has two variables, m_s and ℓ. It is well known that the periodic waves in the sliver are produced due to mechanical reasons in processing. If we assume that the draft from the feed roll to the collecting surface of the rotor is ʹλ (spinning draft), then

$$m_s = a.\cos\left[\frac{2\pi}{n}.\ell\right]$$

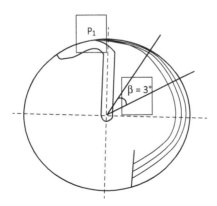

FIGURE 8.5
Fiber layer formation to compose the yarn—β phase angle. (From unknown, 1969.)

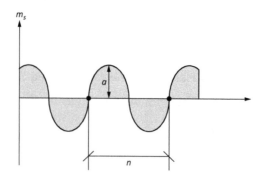

FIGURE 8.6

Distribution of fed sliver mass m_s along its length ℓ (Equation 8.31). (From Soliman H.A., 1980, Lecture notes, TED, Alexandria University, Alexandria, Egypt.)

Legend for Figure 8.6

n = harmonic wave length
a = amplitude of periodic wave
ℓ = general sliver length
m_s = sliver mass distribution (harmonic)

$$'\lambda = \frac{V_c}{V_f}$$

where:
V_c = Surface speed of the collecting surface of the rotor
V_f = Surface speed of the feed roller

Accordingly, the draft $'\lambda$ in Equation 8.31 will have the following form:

$$m_c = \frac{\partial}{'\lambda} . \cos\left[\left(\frac{2\pi}{n'\lambda}\right).\ell\right] \tag{8.32}$$

where:
m_c = The mass of a fiber layer on the rotor collecting surface that is a function of the number of fibers in the cross section

$\dfrac{\partial}{'\lambda}$ = Amplitude of the periodic wave of the fiber layer in the rotor groove

ℓ = Fiber layer length in a general position

This means that Equation 8.32 has two variables, m_c and ℓ.

Legend for Figure 8.7

B = Total number of layers in fresh rotor yarn that equals back doubling

$\dfrac{u}{B}$ = Shift between each two successive layers

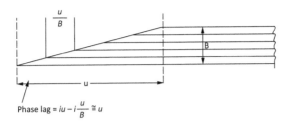

FIGURE 8.7
Formation of fiber layers along rotor circumference U with shift $i\dfrac{U}{B}$. (From Soliman H.A., 1980, Lecture notes, TED, Alexandria University, Alexandria, Egypt.)

$i\dfrac{u}{B}$ = General shift of the layer in a general position

i = Number of any layer in general

m_c = Fiber layer's mass distribution = $\dfrac{a}{'\lambda} \cdot \cos\dfrac{2\pi}{n'\lambda}$

a = $\dfrac{a}{'\lambda}$

= Harmonic wave amplitude in fiber layer

$'\lambda$ = Draft from V_c to V_f = $\dfrac{V_c}{V_f}$

= Spinning draft D_s

V_c = Linear speed of rotor collecting surface

V_f = Feed roll surface speed

n = Harmonic wavelength in the sliver

$n'\lambda$ = Harmonic wavelength in the fiber layers that compose the yarn cross section

$$'m_e = \frac{a}{'\lambda}\left[\cos\left(\frac{2\pi}{\lambda}\right)(\ell - in)\right]$$ (8.33)

Harmonic mass of any fiber layer in a general position

As shown previously, the yarn is formed from accumulated layers, the number of which is equal to back doubling as shown in Figure 8.5 where there is a shift between each two successive layers:

$$= \frac{u}{B}$$

where:
u = Rotor circumference
B = BD

But if the number of fiber layers at a general moment that range from unity to back doubling (B) is i, then

$$1.0 \leq i \leq B \tag{8.34}$$

For the shift between different layers as a phase angle (Figure 8.7):

$$= iu - i\frac{u}{B} \tag{8.35}$$

$$\cong iu$$

In Equation 8.35, we neglected u/B due to the large value of B. By reformatting Equations 8.22 and 8.33, we get

$$'m_c = \frac{a}{'\lambda}\left[\cos\left(\frac{2\pi}{\lambda}\right)(\ell - iu)\right] \tag{8.36}$$

Equation 8.36 gives the mass of any fiber layer or the number of fibers in the cross section of any fiber layers in a general position.

The final mass of the rotor-spun yarn m_y is the summation of $'m_c$ in Equation 8.36, that is,

$$m_y = \sum 'm_c \qquad \text{(Figure 8.8)}$$

$$= \frac{a}{'\lambda}\sum \cos\left[\left(\frac{2\pi}{n'\lambda}\right)(\ell - i\lambda)\right] \tag{8.37}$$

The amplitude R of the wave in Equation 8.37 is

$$R = \frac{a\pi}{u}.\sin\frac{\pi u}{\pi\lambda} \tag{8.38}$$

where λ is the total draft from the sliver to the yarn that is smaller than $'\lambda$.

Legend for Figure 8.8

my = Yarn mass distribution of harmonic wave

$\dfrac{a}{'\lambda}$ = Amplitude of rotor yarn harmonic wave

$n'\lambda$ = Wavelength of harmonic wave in the rotor yarn

$$my = \frac{a}{'\lambda}\sum_{i=1}^{i=B}\cos\frac{2\pi}{n'\lambda}[\ell - i]$$

$$= \sum m_c$$

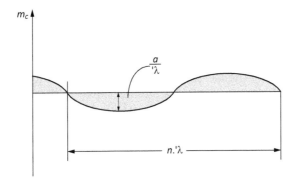

FIGURE 8.8
Harmonic wave in the rotor-spun yarn. (From Soliman, H.A., 1980, Lecture notes, TED, Alexandria University, Alexandria, Egypt.)

The resultant amplitude of the harmonic wave in the rotor yarn (due to back doubling) is

$$R = \frac{a.\pi}{\pi U} \sin \frac{\pi U}{n-1}$$

where the amplitude of the harmonic wave in the rotor yarn without back doubling is P:

$$P = \frac{a}{\lambda}$$

where:
a = Amplitude of harmonic wave in the sliver
λ = Total draft $D_T = V_d / V_f$

The ratio between

$$\frac{R}{P} = Q$$

where:

$$Q = \frac{\lambda}{\pi u} \sin \frac{\pi u}{\lambda}$$

Q = Non-improvement coefficient
$Q' = 1 - Q$

- Improvement coefficient

  $\pi\lambda$ = Periodic/harmonic wavelength on the yarn that is produced from the rotor spinning machine

R   = Amplitude of harmonic wave on the rotor-spun yarn
U   = Rotor circumference
a, n = Wavelength and its amplitude in the fed sliver

If we assume that the fed sliver was fed to a rotor unit (spun box) in the spin position with a total draft $= '\lambda$, that is, equal to the draft from the fed roller to the rotor groove (collecting surface), then the harmonic (periodic) wave in the fed sliver will have a wavelength equal to n and an amplitude equal to a. This wave in the rotor yarn will have a wavelength $n'\lambda$ and an amplitude $\partial / '\lambda$. We will symbolize the amplitude of the harmonic wave in the rotor yarn by P:

$$P = \frac{\partial}{'\lambda} \tag{8.39}$$

Equation 8.39 means that we produced rotor yarn without back doubling. But the amplitude of the harmonic (periodic) wave in the rotor yarn that is produced by back doubling will follow Equation 8.38. Next, it is required to determine the improvement in the rotor yarn due to back doubling; this will be symbolized by Q:

$$Q = \frac{R}{P} \tag{8.40}$$

$$= \frac{\lambda}{\pi u} . \sin\left(\frac{\pi u}{\lambda}\right)$$

where Q is the ratio between amplitudes of harmonic waves in a rotor yarn with R and without P (back doubling).

The value of improvement will be symbolized by $'Q$ where:

$$'Q = 1 - Q \tag{8.41}$$

Equation 8.40 is presented graphically in Figure 8.9.

From Figure 8.9, it is shown that the relative amplitude Q (an improvement) is equal to zero when $\lambda / u = 1/2, 1/3, 1/4$, and so on, that is, when the wavelength of the sliver is $1/2, 1/3, 1/4$, and so on, of the rotor circumference. This means that the improvement $'Q$ will be unity, that is, 100%. According to Figure 8.9, the relative amplitude Q will increase by the increase of the amplitude of the harmonic wave in the sliver from one, two, three, four, or five times the rotor circumference, that is, $'Q$ will be decreased. If this phenomenon reaches sixfold of the rotor circumference, $'Q$ will be zero (no improvement takes place).

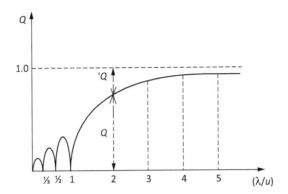

FIGURE 8.9
Relationship between relative amplitude and relative wavelength— harmonic type. (From Soliman H.A., 1980, Lecture notes, TED, Alexandria University, Alexandria, Egypt.)

It can be concluded that the harmonic wave in the fed sliver to the rotor spinning machine can disappear completely when its wavelength ranges from 1/2 to 1.0 of rotor πD_R, but when the periodic wavelength in the fed sliver is equal to or greater than six times the rotor's circumference, the improvement is canceled completely.

Legend for Figure 8.9

$$P* = \frac{a}{'\lambda}$$

$$Q = \frac{R}{P*} = \frac{n.\lambda.}{\pi n} \sin \frac{\pi u}{n.\lambda}$$

$$= \frac{\lambda}{\pi u} \sin \frac{\pi u}{\lambda}$$

= Relative amplitude
= Reduction in the amplitude due to back doubling

$$= \frac{\text{Resulting amplitude with BD}}{\text{Actual periodical amplitude in ring spinning}}$$

8.4 Summary Points

1. The basic definition of back doubling is that it is the number of fiber layers from the final rotor yarn cross section that is formed mathematically; it can be defined as

$$BD = \frac{1}{c}$$

$$= \frac{V_c}{V_d} \qquad \text{(a)}$$

where:
c = Fiber condensation in the rotor
V_c = Linear speed of the fibers' collecting surface
V_d = Yarn delivery speed

Several subdefinitions can be written as follows:

$$BD = \pi D_R . T - 1$$

$$\cong \pi D_R . T \qquad \text{(b)}$$

where:
BD = Back doubling
π = Rotor diameter
T = tpm or tpi according to the units of the rotor

$$BD \times D_T = D_S \qquad \text{(c)}$$

where:
BD = Back doubling
D_T = Total draft = V_d / V_f
D_S = Spinning draft = V_c / V_f
V_d = Delivery speed
V_c = Surface speed of rotor groove
V_f = Fed speed of feed roller

2. The definitions a, b, and c are valid for the back-doubling phenomenon, as the fiber layer that slides from the rotor sliding wall to its
groove (collecting surface) and the successive deposits of fiber layers compose the total number of fibers in a rotor yarn (fresh) cross
section.

3. The back doubling in the rotor groove is responsible for the leveling (canceling) of the random waves in the sliver and is due to toothed drafting.

4. The periodic (harmonic) waves in the fed sliver due to mechanical faults in the previous processes can be canceled completely if the length of the harmonic wave is equal to $1/4, 1/3, 1/2$, or 1.0 of the rotor circumference where the improvement coefficient Q' is equal to 1.0 (100%). When the harmonic wavelength is equal to two, three, four, or five turns of the rotor circumference, the improvement coefficient Q' decreases gradually to zero; when the periodic wavelength in the fed sliver equations is more than sixfolds of the rotor circumference, that is, when

$$\lambda / u \geq 6.\pi D_R \tag{d}$$

but when

$$\frac{\lambda}{u} = \left[\frac{1}{4}, \frac{1}{3}, \frac{1}{2}, \text{and} 1\right].\pi D_R \tag{c}$$

a complete canceling of the harmonic wave in the fed sliver can take place.

5. The peeling-off (sweeping) point is the absolute point of connection between fibers on the collecting rotor surface and the open-end yarn, which combine to form a fresh rotor yarn.

6. The peeling-off point rotates and rolls on the rotor surface in the same rotational direction of the rotor with a relative velocity of $'N_p$:

$$'N_p = \frac{V_d}{\pi D_R} \tag{d}$$

where:
$'N_p$ = Relative speed of point P with respect to the rotor
V_d = Delivery speed of yarn
D_R = Rotor's diameter

7. The absolute velocity of point P is N_p:

$$N_p = 'N_p + N_R$$

where:
N_R = Rotor speed in RPM
N_p = Absolute point P's speed in RPM

8. The accurate tpi or tpm (T) is

$$T = \frac{N_p}{V_d} \qquad \text{(e)}$$

But, approximately,

$$T\left(\text{tpm or tpi}\right) = \frac{N_R}{V_d}$$

where:
N_p = Point P's absolute velocity in RPM
N_R = Rotor speed in RPM
$'N_p$ = Relative speed of point P

9. Whether a sliver's harmonic wave can be canceled under certain conditions depends on the improvement coefficient Q':

$$Q' = 1 - Q \qquad \text{(f)}$$

where:
Q' = Improvement coefficient
Q = Non-improvement coefficient

$$Q = \frac{\lambda}{\pi u}\sin\left(\frac{\pi u}{\lambda}\right) \qquad \text{(g)}$$

where Q is the relative amplitude reduction in harmonic wave amplitude due to back doubling.

Review Questions

Q#1: Mention with mathematical formulae the different definitions of the back-doubling phenomenon.

Q#2: What is the basic equation of back doubling inside the rotor spinning machine's rotor?

Q#3: Prove mathematically that BD is

$$BD = \pi D_R . T - 1$$

$$BD. D_T = D_s$$

$$BD = N_R / 'N_p$$

where:
BD = Back doubling of fiber layers inside the rotor
D_R = Rotor's diameter
T = Twists (turns) per unit length (tpm or tpi)
D_T = Total draft

$$= V_d / V_f$$

where:
V_d = Yarn speed
V_f = Feed speed
D_s = Spinning draft

$$= \frac{V_c}{V_f}$$

where:
V_c = Linear speed of rotor's collecting surface
$'N_p$ = Relative speed of point P's connection between the open end of fresh rotor yarn and the fiber layer on the rotor groove

Q#4: Write qualitatively only about the relationship between the harmonic wave in the fed sliver of the rotor box position and back doubling in the rotor.

Answers to Review Questions

Q#1: See the three definitions of back doubling in Section 8.3.1.
Q#2: See Equation 8.20.
Q#3:
- See first definition and Equations 8.21 through 8.23.
- See second definition and Equations 8.24 through 8.29.
- See Equations 8.25 through 8.28

Q#4: See Section 8.3 on back doubling and harmonic waves in the fed sliver to the rotor box. Answer descriptively (i.e., without equations).

Review Problems

Q#1: A rotor spin box produces a rotor yarn from cotton fibers with $N_m = 20$ and mtex $= 200$. If the rotor $D_R = 65$ mm and $N_R = 30$ kRPM, explain with the aid of an illustration the back doubling that takes place in the rotor.

Q#2: A rotor spinning machine has a rotor diameter of 40 mm and runs at a speed of 100 m/min. If the rotor rotates at its maximum possible speed (h):

$$N_R = \frac{3.1 \times 10^3}{D_r(\text{mm})} \text{ RPM (h)}$$

and the produced yarn count $N_e = 30$. What will be the back doubling? Use different formulae.

Q#3: For a cosine wave (harmonic wave/periodic wave) in the fed sliver to a rotor unit, what will be

- The final mass of rotor yarn m_y
- The amplitude of the rotor yarn's wave R
- The improvement coefficient $\dot{\varphi}$ of the rotor yarn wave's amplitude

Q#4: Represent graphically the relationship between non-improvement coefficient Q and the ratio λ/u ($\lambda =$ total draft and $u =$ rotor circumference).

Answers to Review Problems

Q#1: See Active Example 2.

Q#2: By using the Shlafhorst formula,

$$N_R = \frac{3.1 \times 10^6}{0.040}$$

$$= 77,500 \text{ RPM}$$

Delivery speed V_d is

$$V_d = 100 \text{ m/min}$$

$$\therefore \text{tpm} = \frac{77,500}{100}$$

$$= 775$$

Rotor circumference is

$$\pi D_R = \pi \times 0.04$$

$$= 0.1256 \text{ m}$$

$$\therefore \text{BD} = \pi D_R \text{ tpm}$$

$$= 0.1256 \times 775 - 1$$

$$= 97.34$$

$$\cong 97.0$$

Another solution:

$$V_c = \pi D_R . N_R$$

$$= \pi \times 0.04 \times 77,500$$

$$= 97.34 \text{ m/min}$$

$$\cong 97.0$$

$$\text{BD} = \frac{V_c}{V_d}$$

$$= \frac{9734}{100}$$

$$= 97.34$$

$$\cong 97$$

Q#3:

- ($\rightarrow$ Formula 8.36)
- ($\rightarrow$ Formula 8.37)
- ($\rightarrow$ Formulae 8.40 and 8.41)

Q#4: See Figure 8.9.

Bibliography

Elhawary I.A., 2008, The technology of the rotor spinning machine, Lecture notes, TED, Alexandria University, Alexandria, Egypt.

El Moghazy Y. and Chewning, C.H., Jr. 2001, *Cotton Fiber to Yarn Manufacturing Technology*, Cotton Inc., Cary, NC.

Rohlena V., 1974, *Open-End Spinning*, Technical Literature Publisher, Prague, Czech Republic.

References

Soliman H.A., 1980, Lecture notes, TED, Alexandria University, Alexandria, Egypt.

Stadler H.W., 1975, *Influence of Rotor Speed on the Yarn Manufacturing Process*, Textile Trade Press, Manchester, UK.

Appendix I

Section 1: Technology of Rotor Spinning

Rotor spinning is one type of open-end spinning techniques that is applied widely in cotton spinning (McCreight et al., 1977; El Moghazy and Chewning, 2001) and to some extent in wool spinning. Rotor spinning involves the individualization of cotton fibers by an opening device and the assembling and twisting of the fibers in a rotor as shown in Figure A1.1. In an actual rotor spinning machine, a second drawing frame sliver is presented to a feed roller (Elhawary, 2014). The fibers of the fed sliver are then opened to the level of the individual fibers by a combing (opening) roller covered by a saw-toothed wire cloth or pinned cloth. Once opened, the cotton fibers pass through a transport duct where they are separated further (air draft) before being deposited on the rotor collecting surface. The generated inertia forces—by the rotor rotating at high velocities—make the fibers assemble and compact on the rotor collecting surface, forming a ring. The fiber ring is then swept from the rotor by a newly produced yarn end that has untwisted fibers. For every one rotor turn, one twist is inserted, changing the fiber bundle into yarn as it is taken out of the rotor via the navel. The produced rotor yarn is taken up onto an across-wound package, canceling the necessity for a separate winding operation.

Figure A1.2 illustrates the comparative structure of rotor-spun yarns with large and small rotor diameters. The presence of bridging/belting fibers is less common with smaller rotors than with larger rotors. The typical end uses of rotor yarn are: sweaters, denim, upholstery, t-shirts, blankets, towels, underwear, sheeting, and so on. The advantages of rotor-spun yarn include the following: (1) lower yarn imperfections, especially avoiding long thick and thin areas; (2) good appearance of unit fabrics; (3) superior dye ability; (4) less torque with respect to ring-spun yarn; and (5) sophisticated real-time quality and production monitoring on each spin box.

The disadvantages of rotor-spun yarns are: (1) low yarn tenacity (70%–80% of ring-spun yarn); (2) high pilling (propensity); (3) knitting needles are wearable compared to ring-spun yarn; and (4) the rotor yarn structure is identical compared to the ring-spun yarn as shown in Figure A1.3. The following cotton fiber properties are important for rotor-spun yarn production: fiber strength, micronaire fineness, fiber length, and short fiber content. The important properties of synthetic fibers with respect to rotor-spun yarn are: fiber denier, fiber length, fiber tenacity, and fiber finish. The Engineered Upgraded Yarn Quality Factor (EUG-YQF) was introduced at the Belt Wide Cotton Conference (BWCC) in the United States (Elhawary, 2015). The cotton-fed sliver for rotor spinning mechanics has the following critical properties: (1) sliver regularity

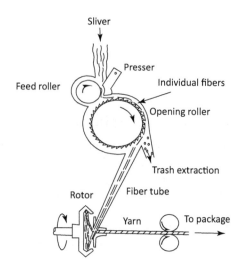

FIGURE A1.1
Schematic representation of the rotor spinning process. (From McCreight D.J., Feil R.W., Booter Baugh J.H., and Backe E.E., 1997, *Short Staple Yarn Manufacturing*, Carolina Academic Press, Durham, NC.)

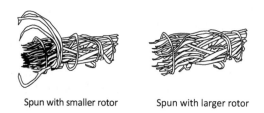

Spun with smaller rotor   Spun with larger rotor

FIGURE A1.2
Illustration of the comparative structures of rotor yarns spun with large and small diameter rotors. (From McCreight D.J., Feil R.W., Booter Baugh J.H., and Backe E.E., 1997, *Short Staple Yarn Manufacturing*, Carolina Academic Press, Durham, NC.)

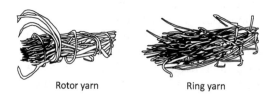

Rotor yarn   Ring yarn

FIGURE A1.3
Illustration of the ring and rotor spun yarns. (From McCreight D.J., Feil R.W., Booter Baugh J.H., and Backe E.E., 1997, *Short Staple Yarn Manufacturing*, Carolina Academic Press, Durham, NC.)

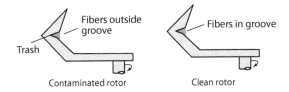

FIGURE A1.4
Schematic comparison of a clean rotor to a rotor contaminated with trash buildup. (From McCreight D.J., Feil R.W., Booter Baugh J.H., and Backe E.E., 1997, *Short Staple Yarn Manufacturing*, Carolina Academic Press, Durham, NC.)

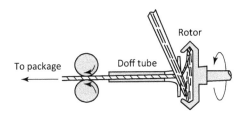

FIGURE A1.5
Simplified schematic illustration of the doff tube as used in a rotor. (From McCreight D.J., Feil R.W., Booter Baugh J.H., and Backe E.E., 1997, *Short Staple Yarn Manufacturing*, Carolina Academic Press, Durham, NC.)

(mid-term and short term); (2) fiber straightening and alignment; and (3) sliver slubs. Dust or trash that remains in the sliver can potentially accumulate in the rotor groove (collecting surface) and prevent the yarn from being created appropriately as in Figure A1.4.

The doff tube shown in Figure A1.5 is placed behind the navel and provides a guide for the yarn to be taken out of the spin box. In modern rotor spinning machines, a new modification to the doff tube is made by the addition of a twist trap or torque twist as shown in Figure A1.6. The torque stop increases the false twist effect in the rotor groove and consequently contributes to a higher rotor RPM without an increase in end breaks. Modern rotor spinning machines are provided with an auto piecer that travels along the machine and automatically rejoins the broken ends.

Summary Points

1. The number of fibers per cross section of yarn can be estimated by the formula:

$$\text{Fibers per cross section} = \frac{15,000}{[\text{Mic}][\text{Yarn Ne}]} \tag{A1.1}$$

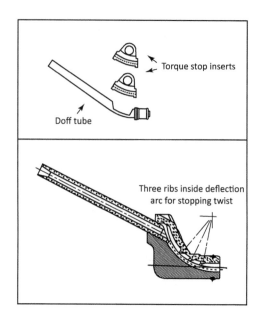

FIGURE A1.6
Two schematic views of the torque stop or twist trap device. (From McCreight D.J., Feil R.W., Booter Baugh J.H., and Backe, E.E., 1997, *Short Staple Yarn Manufacturing*, Carolina Academic Press, Durham, NC.)

where:
 Mic = Microgram per inch (fineness) of cotton fiber
 Ne = English yarn count

$$\text{Fibers per cross section} = \frac{5315}{[\text{Denier}][\text{Yarn Ne}]} \tag{A1.2}$$

where:
 Denier = Artificial fiber denier
 Ne = English yarn count

For cotton spinning:

$$D_R \geq 1.2 * SL \tag{A1.3}$$

where:
 D_R = Rotor diameter
 SL = Cotton fiber staple length

For synthetic fibers (artificial fibers):

$$D_R \geq 0.83 * FL \tag{A1.4}$$

where FL is the fiber length.

3. On rotor spinning machines, the end breaks are between 100 and 200 per 1000 rotor hours and quality cuts should be less than 100 total per 1000 rotor hours.

4. The critical factors in the combing roll zone for optimum quality are combing roll metallic cloth conditions, choice of combing roll wire type, speed of the combing rolls, setting between combing roll and feeding tray, and feed clutch wearability. Most spinning mills set the combing roll velocity between 7600 and 8600 RPM.

5. Through the development of new bearings, smaller rotors with better fiber control mean rotor speeds of up to 150 kRPM are now possible.

6. The formation of wrapper (belted–bridge) fibers depends on the diameter and speed of the rotor, navel performance, internal friction between the fibers and rotor collecting surface, and fiber length.

7. The EUG-YQF of the rotor-spun yarn is calculated by the formula (Elhawary, 2010):

$$E = \frac{CSP*\text{Yarn tenacity in MPa} \left(\text{mega pascal} \right)}{U\%} \qquad (A1.5)$$

where:

$$\begin{aligned} CSP \quad &= \quad \text{Count strength product} \\ U\% \quad &= \quad \text{Mean percent deviation as measured by Uster tester} \\ \text{Yarn tenacity} \quad &= \quad \text{Single end strength in MPa} \end{aligned}$$

8. Nowadays, the smallest rotor diameter is 28 mm and this can attain 160 kRPM.

9. The combing roller's (opening beater and opening cylinder) opening action is associated with the cleaning of cotton fibers (Elhawary, 2010).

10. The productivity of a rotor spinning position is 10 or more times that of a ring spinning position (McCreight et al., 1997).

11. The regularity (mass variation CVm%) of the rotor-spun yarn is better than that of the ring-spun yarn due to the back-doubling phenomenon inside the rotor during yarn formation, that is, individual fibers building up on the rotor collecting surface (Elhawary, 2010).

12. Wool is usually spun on the rotor spinning system. Despite progress in this trend, 42 tex to 111 tex yarns are spun successfully from 20.5 μm wool at a delivery speed of 100 m/min. The spinning problems are residual grease level (0.10%–0.3%) and fiber length. For example, wool fiber of mean length = 30–40 mm is required for a rotor diameter of 64 mm with the longest 1% of fibers not exceeding 60 mm (Simpson and Crawshaw, 2002).

Review Questions

Q#1: Write a formula for each of the following:

(1) The number of fibers per rotor yarn cross section; (2) the rotor-spun yarn quality factor; (3) the relationship between rotor diameter and fiber length for both cotton and synthetic yarns; (4) the turns per meter (tpm) of a rotor-spun yarn using the rotor speed and the delivery (take-up) speed; and (5) the total draft of the rotor spinning machine.

Q#2: Explain the phenomenon of wrapper (belted-bridged) formation in rotor spinning.

Q#3: What are the advantages and disadvantages of rotor-spun yarn?

Q#4: Make a line diagram for the rotor spinning process. Label the different parts.

Q#5: Using freehand sketches only, show (1) the difference in rotor yarn structure of small rotor diameters and large rotor diameters; and (2) ring-spun yarn structure and rotor-spun structure.

Q#6: Using a line diagram, show the trash effect on yarn formation inside the rotor.

Q#7: Compare graphically: (1) different wiring cloths of the combing roller; (2) different doffing tube navels; (3) an S-, T-, U-, G-, and K-rotor's profile; and (4) different rotors' grooves.

Q#8: What is meant by torque stop in the rotor-spun yarn? How does it affect the process of yarn formation?

Answers to Review Questions

Q#1:

a. See Equation A1.1 and divide the result by 2.
b. See Equation A1.5.
c. See Equation A1.3.
d. tpm = n_R / V_d, where n_R = rotor RPM and V_d = take-up speed in meters per minute.
e. Total draft = V_d / V_f, where V_f = surface speed of feed roller or = sliver ktex/yarn tex.

Q#2: Once each revolution of the yarn peeling-off point from the collecting surface is completed, the fibers entering the rotor from the transport duct can then be integrated with the yarn, that is, some of the entering fibers can be incorporated into the yarn body.

Q#3: For the advantages, see Summary Points 11 and 12. The main disadvantage is the low yarn strength.

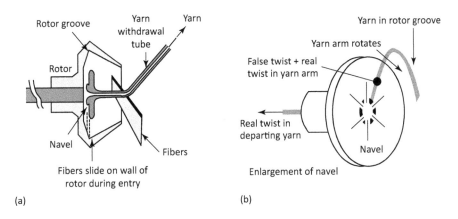

FIGURE A1.7
Fiber and yarn in the rotor. (From Lord P.R., 2012, *Handbook of Yarn Production*, Woodhead Publishing, Cambridge, UK.)

Q#4: See Figure A1.1.

Q#5: See Figure A1.3.

Q#6: See Figure A1.4.

Q#7:

    a. See Appendix A1 Section 3: Opening Device (Combing Roller) (SKF, 1991).

    b. See Appendix A1 Section 2: The Rotor, Doffing Tube and Navel, and Fiber Transport Channel. See Figures A1.8 and A1.10.

Q#8: See Figure A1.6. The torque stop increases the false twist in the rotor groove and consequently contributes to a higher rotor RPM.

Section 2: The Rotor, Doffing Tube and Navel, and Fiber Transport Channel

The rotor (cup) configuration (profile) with its associated elements (doffing tube with navel and fiber draft) is illustrated in Figure A1.7.

The transported opened (drafted) fibers enter the rotor zone to be deposited on the rotor's internal wall where they move to the collecting surface (rotor's groove) to make the fibers ring.

Different rotor groove types are available to create flexibility in yarn bulk. Different sizes and types of rotor groove are shown in Figure A1.8. The S-groove and U-groove rotor is used for high bulk yarns, whereas a G-groove rotor is applied to produce more yarn. The F-groove rotors are used to provide the leanest, least hairy, and strongest rotor-spun yarn. K-grooves have been applied in rotors with small diameters.

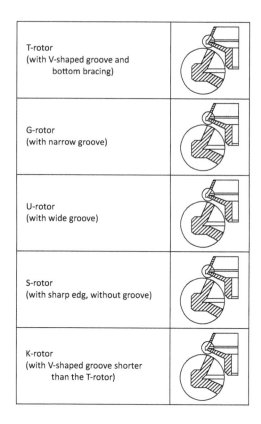

T-rotor (with V-shaped groove and bottom bracing)	
G-rotor (with narrow groove)	
U-rotor (with wide groove)	
S-rotor (with sharp edg, without groove)	
K-rotor (with V-shaped groove shorter than the T-rotor)	

FIGURE A1.8
Different rotor types applied in rotor spinning machines. (From McCreight D.J., Feil R.W., Booter Baugh J.H., and Backe, E.E., 1997, *Short Staple Yarn Manufacturing,* Carolina Academic Press, Durham, NC.)

The rotor speed significantly affects the rotor yarn properties and wrapper fiber formation, but with the suitable selection of rotor and transport duct profiles (design) and fiber, a high rotor speed of up to 150 kRPM can be worked to achieve maximum rotor velocities where a rotor of small diameter is applied to create an inertia force (centrifugal force) on the ring of fibers in the rotor groove to maintain control. The maximum allowable surface speed is 220 m/s to keep the centrifugal force under control. For a rotor diameter = 33 mm, the RPM = 127 kRPM, for first-generation rotors where rotor diameters = 56 mm, then the RPM = 75 kRPM, while for rotor diameter = 46 mm, the RPM = 91 kRPM. Of course, first-generation rotors could not run with such high speed due to the lack of bearings in rotor spindle technology. At the first error of rotor spinning like BD700 M 69, the maximum rotor surface speed was about 90 m/s. For rotors with a diameter of 56 mm, the maximum speed was 31 kRPM. A guide for selecting a suitable rotor type and size for a certain yarn count is shown in Figure A1.8. The continuous

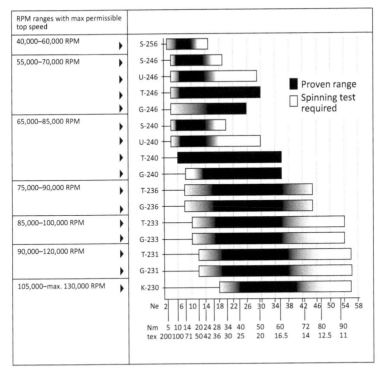

FIGURE A1.9
Guide for choosing a suitable rotor size and type for a certain yarn count. (From McCreight D.J., Feil R.W., Booter Baugh J.H., and Backe, E.E., 1997, *Short Staple Yarn Manufacturing*, Carolina Academic Press, Durham, NC.)

contact between the cup (rotor) stem with the thrust bearing can generate wear on the rotor stem (a concave wear pattern) that can cause the axial location of the cup (rotor) to change course, meaning the transported fibers to the rotor will not be fed correctly. The yarn quality will then deteriorate. McCreight et al. (1997) have written that the CVm% of 24 s yarn for a worn rotor stem is 16.7%, while for a new rotor it is 15.9%, indicating a 1% loss. The stem problems of the rotor were canceled for air thrust bearing, as licensed by Rieter Wintertur, Switzerland.

For the vertical axis, delete the digit 2 to determine the rotor diameter in millimeters. McCreight et al. (1997) stated that the rotor's cup begins to wear after 20,000 h of running, or when the yarn CVm% is increasing by 1% more than the new cup (rotor). Also, the test for the moiré effect is a good indicator.

The modern drive or bearing of a highly rotated small diameter rotor is a Twin Disc type. These discs are covered by rubber rings that meet the rotor spindle. When they are damaged, the quiet and smooth running of the rotor is impossible and the replacement of the rubber rings is required. The wax and trash of cotton fibers that build up on the rubber rings can affect the yarn

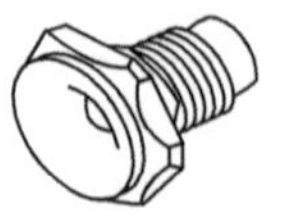	Smooth navel without grooves to spin non-hairy yarns of cotton and acrylics.
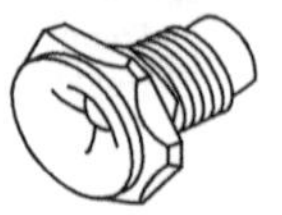	3-groove navel for low yarn hairiness with all raw materials.
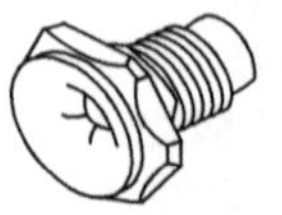	4-groove navel for low yarn bulk with all raw materials.
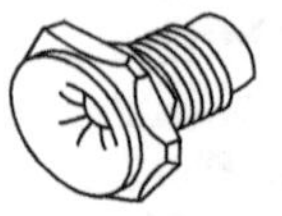	8-groove navel for use with smaller diameter rotors for low twist yarns.
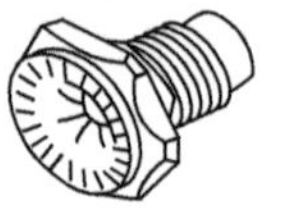	8-groove navel with knurled rim for high hairiness and high bulk in yarns with all raw materials.
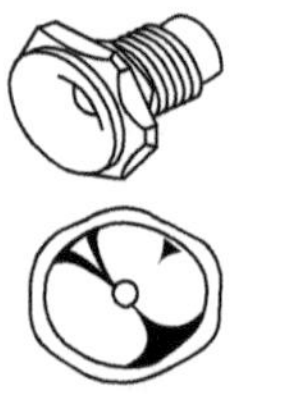	Smooth navel with fluted insert in navel throat for greater yarn hairiness and bulk with materials with low short fiber content.
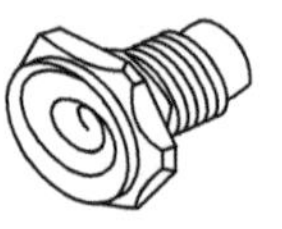	Ceramic navel with a springly arranged ridge for low hairiness and low bulk in 100% cotton yarns.

FIGURE A1.10

General engineering views of various navel types in rotor spinning. (From McCreight D.J., Feil R.W., Booter Baugh J.H., and Backe, E.E., 1997, *Short Staple Yarn Manufacturing*, Carolina Academic Press, Durham, NC).

quality. The rubber rings need to be cleaned periodically every six months (preventative maintenance).

The Twin Disc bearing or drive is explained in detail in Chapter 3. The antifriction bearings of the Twin Discs should be checked using smooth movement every six months.

For any type of rotor bearing or drive, the tangential belt drive is essential (except for the motor of each rotor). Belt wear causes unstable and abnormal rotor operation that leads to end breaks. The noise level in decibels (dBA) of a bad belt drive is a good measure as well as a 50% increase in end breaks.

The Doffing Tube and Navel

The fresh yarn segment inside the rotor is withdrawn via a navel and doff tube (Figure A1.7b). These two elements affect both the yarn quality and spinning performance.

Both the navel and doff tube generate friction on the yarn during its withdrawal, which generates a false twist effect where the yarn arm strength increases. Different navels with various friction surfaces are available. Most navels are manufactured from ceramic but others are made from steel for artificial fibers to avoid the bad effects of heat generation.

Navels have grooves or ridges (cut or molded) on the navel orifice (0-, 3-, 4-, or 8-groove). A1.10 shows a general view of the navel where the deflection of the doff tube is reduced from 90° to 30°–37° for a new rotor machine to reduce yarn withdrawal tension during spinning. Also, a twist trap is incorporated in the doff tube as a torque on Schlafhorst machines. It looks like "speed bumps" on a roadway.

Fiber Transport Duct

The fiber transport duct is a transport element for the opened fibers from the combing roller to the rotor sliding wall (inclined surface). The critical functions for the transport channel are seal condition, channel (duct) wear, and duct alignment. The seal is used on the split ducts to prevent air leaks. As the seal deteriorates, air will leak out of the duct and fiber flow will be highly disturbed, which will deteriorate yarn quality. Also, the rotor seal due to the opening and closing of the spin box can cause the seal to become worn and may cause misalignment between both parts of the transport duct. In addition, the friction between the fibers and duct will create grooves that lead to a deterioration in yarn quality.

Summary Points

1. The rotor of the rotor spinning machine has different names: rotor, pot, cup, and camera.
2. The fiber transport duct or channel is a transport element for the opened fibers from the combing roll to the rotor sliding wall (inclined wall). The channel has a wide opening toward the feed rolls and

narrow opening toward the rotor to accelerate the fiber flow. The air speed ratio between both of the openings is 10:1, that is, toward the rotor, the air speed is 10 times greater at the transport channel inlet. The seal between both the upper and lower parts of the duct is to prevent air leakage to maintain yarn quality.

3. The critical factors for the transport channel are: seal condition, channel wear, and duct alignment. The seal is used on the split channel to prevent air leaks. As the seal deteriorates, air will leak out, which affects fiber flow and consequently lowers yarn quality.

4. The delivered fibers on the rotor slide on the conical rotor wall can be straightened to a certain extent under the centrifugal force effect during their movement downward to the collecting surface.

5. The deposited individual fibers form layers that accumulate to form the final yarn and back doubling occurs which improves the yarn CVm%.

6. The part of the yarn between the navel and the peeling-off point on the collecting surface is named the "yarn arm" where it forms a balloon inside the rotor. Mostly, it rotates positively, that is, in the direction of the rotor revolutions. The rotor grooves are highly connected with yarn quality and there are several different types of grooves.

7. The navels at the lower end of the doffing tube play an important role in yarn formation and quality. There are different shapes of navels.

8. The angle of contact between the yarn arm and the navel at the lower end of the doffing tube ranges in practice from 30° to 70° to create a suitable yarn tension for the withdrawn yarn via the doffing tube.

9. The speed of yarn winding on the winding head of the rotor spinning machine is much slower than the surface speed of the rotor collecting surface so that the fiber has the chance to be condensed properly (effective back doubling + required number of fibers per yarn cross section).

10. The surface speed of the rotor is calculated by the formula:

$$v_R = \pi d_R \cdot N_R \times 1.67 * (E-5) \text{ m/s} \tag{A1.6}$$

where:

v_R = Rotor surface speed in meters per second
d_R = Rotor diameter in millimeters
N_R = Rotor RPM

Nowadays, the maximum allowable surface speed is 220 m/s. For first-generation rotors, the maximum allowable speed was 90 m/s. As the rotor diameter decreased from 56 to 28 mm, the rotor RPM increased from 30 to 150 kRPM.

11. The rotor spinning machine winding head is the same as the winding machine head. The critical factors for the winding zone are angle of wind, wax application, cradle pressure, yarn tension setting, and so on.

12. To keep the centrifugal force constant on the built fibers on the collecting surface, irrespective of the rotor diameter and RPM, then $N_R \cdot d_R$ = constant, that is, the decrease of rotor diameter means an increase in RPM to achieve a constant surface speed.

Review Questions

Q#1: What are the different names for
   a. Rotor
   b. Combing roller

Q#2: What is the function of the transport fiber duct?

Q#3: What is the role of the yarn doffing tube?

Q#4: Write a formula for calculating the surface speed of the rotor.

Q#5: Describe the different shapes of the rotor grooves.

Q#6: What are the different types of navels?

Q#7: What is meant by back doubling in the rotor?

Q#8: What do you know about ballooning inside the rotor?

Answers to Review Questions

Q#1:
   a. Rotor = cup/pot/camera
   b. Combing roller = opening device/beater spindle/opening cylinder

Q#2: To transport opened (detached) fibers from the surface of the combing roller to the rotor (conical wall).

Q#3: To guide the fresh segment under a specified tension from inside the rotor to the winding head.

Q#4:

$$v_R = \pi d_R \cdot N_R \times 1.67 * (E - 5) \text{m/s}$$

where:
   v_R = Surface speed of roller in meters per second
   N_R = Rotor RPM
   d_R = Rotor diameter in millimeters

Q#5: See Figure A1.8.

Q#6: See Figure A1.9.

Q#7: Back doubling in the rotor groove refers to fibers building up in adjacent layers to form the required number of fibers per cross section of the rotor yarn while at the same time improving the yarn CVm%.

Q#8: The yarn arm (between the navel and the peeling-off point on the collecting surface of the rotor) when subjected to centrifugal force due to the rotor RPM will become curved, taking on a balloon shape.

Section 3: The Opening Device (Combing Roller)

The opening device includes the opening cylinder, combing roll, opening roll, and so on.

The combing roll opens the fed sliver in much the same way as makes a licker in a cotton card. The fibers are separated from the fed sliver as shown in Figure A1.11. The detached (separated) fibers are directed to the rotor via a transport duct.

The opening cylinder can be clothed with needles (metallic wire) that are highly applicable in practice despite pinned cloth having practical opening and wear capabilities (McCreight et al., 1997).

However, hardened and ground metallic card cloth (MCC) wires with a surface diamond dust embedded in nickel have been developed. Other sharp edges subjected to damage are usually treated with wear-resistant matters (McCreight et al., 1997). Metallic wire cloth is required for cotton and its blends that are produced with high output, while pinned cloth for the

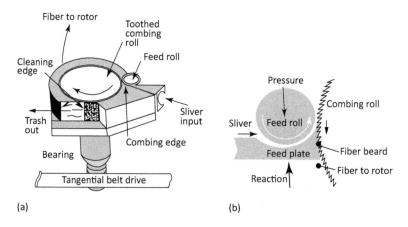

(a) (b)

FIGURE A1.11

Rotor spinning opening device. (From Lord P.R., 2012, *Handbook of Yarn Production*, Woodhead Publishing, Cambridge, UK.)

opening cylinder is recommended for fragile fibers such as rayon and acrylic that are run at average output values.

The metallic pinned roll cloth of the combing roller can become damaged due to its serviceability in continuous friction with the fed fibers, especially cotton and wool with residual grease where they become entangled with impurities. This can decrease the wear ability of the combing roll teeth or pins. The lifespan of the opening cylinder clothing is finite. The use of fibers with an abrasive finish or dusty fibers increases the value of the wear rate for clothing heat or pins. Therefore, not only are the heat surfaces hardened, but also a surface treatment is applied to enhance their abrasion resistance. As with MCCs, the teeth body has to be toughened to stop brittleness and this can cause the teeth to become bent, which negatively affects the yarn mass variation CVm% and long-term variation.

The antifriction bearings of the opening cylinder can enter a phase of damage due to a long service life. A slip in the tangential belt drive of the combing rolls leads to overheating of the bearings, which makes the bearings' lubricant (grease) lose its viscosity with a large probability of leaking out and evaporating into the mill atmosphere, causing air pollution. By this time, dry friction may occur, that is, direct metallic contact between the rolling elements (balls) and both of the inner and outer races of the bearing. In such a phase, other indentations and cracks are made on both the races and rolling elements. This phase can cause spike energy that can be detected by IRD Mechanalysis instruments via a pickup fixed to the bearing housing. Also, the noise level and power requirements increase as a result of wear or damage to the rolling elements. Approximately the same effect can be attained due to a high rate of material feeding (overloading shock) (see Figure A1.12).

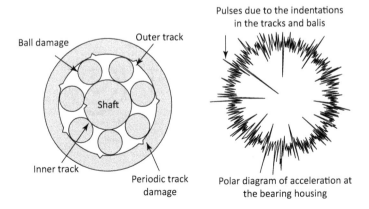

FIGURE A1.12
Damaged ball bearing of beater spindle (combing roll). (From Lord P.R., 2012, *Handbook of Yarn Production*, Woodhead Publishing, Cambridge, UK.)

Most beater spindles (combing rolls) have the following properties (SKF, 1991):

- Wharve diameter d = 23.5 mm
- Combing roll diameter D = 64.6 or 74.6 mm
- Width of combing roll A = 24 or 27.5
- Total length of combing roll spindle L = 91.5 or 93 mm
- Types of metallic cloth OK36, OK37, and OK40, with or without a spiral; selection depends on fiber type (see Figure A1.13)

The choice of the combing roll speed depends on several factors: coarse yarn counts need higher speed; for cotton with a low micronaire value, a high combing roll speed is required and more trash contact in cotton at

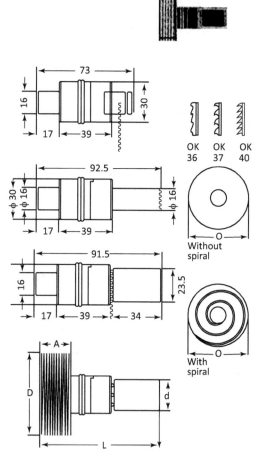

FIGURE A1.13
Combing rolls of rotor spinning machine with exploded views: The choice of combing. (From SKF Group, 2013, Roller Bearings Catalogue, Sweden.)

a high roll speed is achieved. Most plants set the beater spindle (opening cylinder speed) between 7600 and 8600 RPM.

The combing roll wire usually works for loads of 6800–9100 kg, that is, $\cong$7 t (metric tons) and 9 t, depending on the trash content in the cotton or additives to the artificial fibers.

It is important to note the following:

Most wires on the combing roll used in plants are saw-tooth shaped and hardened with diamond and nickel.

Summary Points

1. The opening device contains the combing roller, opening roller, beater, and spindle.
2. The combing roller can be clothed by saw-toothed cloth or pinned cloth, of the German company Burkhardt type. Most industrial rotor spinning machines are equipped with a saw-toothed cloth (metallic wire cloth). There are different types of this metallic wire cloth depending on the type of fed fibers (natural or artificial) and their inherent characteristics. In practice, the combing roll speed ranges from approximately 8.0 to 9.0 kRPM, depending on the type of fiber, trash content, and additives of artificial fibers. The combing roll wire is hardened with diamond and nickel. The change of the combing roll cloth usually takes place after processing 7–9 t, depending on many factors such as type of fiber, roll speed, and so on.
3. The combing roller and rotor are driven by a tangential belt drive that extends over the full length (single side) of the rotor spinning machine.
4. The belt drive of the combing rollers generates heat due to friction with the wharves (driving pulley) of the opening roller; this heat with fed fibers (overfeed) and shock can make the antifriction bearing of the beater (combing roll) spindle enter a damaged phase that can cause spike energy, unsmooth running, noise, and greater energy consumption.
   a. The failure of the bearings can be detected through spike energy evaluation by the use of a pickup (transducer) with special instruments from IRD Mechanalysis.

Review Questions

Q#1: What is the function of the combing roller?

Q#2: Differentiate between a saw-toothed combing roller cloth and its pinned cloth (Burkhardt type).

Q#3: What are the causes of combing roll antifriction bearing damage?

Q#4: What are the functions of air in the transport duct?

Answers to Review Questions

Q#1: To open, detach (separate) the fibers from the fed sliver using a trumpet and feed plate (spring loaded) and by rotating the feed roll. The opened fiber is seized by the combing roller wire cloth to be transferred to the fiber transport duct via air stripping of the fibers from the combing roller cloth.

Q#2: The metallic combing roll cloth is now industrially applied with great success, with certain modifications made to suit different types of fibers with different characteristics. In first-generation machines, it was preferable to use pinned cloth to cover the combing roller outer surface with the aid of certain clips. This type of cloth was used for artificial fibers with a low running speed.

Q#3: The three main reasons can be summarized as follows:

   a. The heat generated from the tangential belt drive due to friction

   b. Opening an overfed fiber besides detaching (separating them) it from the second drawn sliver

   c. The combing roll dynamic unbalance due to mechanical and technological causes

Q#4: A vacuum is created at the end of the rotor spinning machine to generate suction at each spinning box. The negative pressure (suction) removes the fibers from the combing roll metallic cloth and moves them via the transport tube to the rotor. The air carrying the fiber is accelerated by about 50% due to the duct's conical shape in addition to the negative air pressure inside the rotor. The fibers and air acceleration contribute to more fiber individualizations and strengthening that will lead to increasing the fiber extent in the rotor-spun yarn. It is the projection of the fiber length on an axis (usually the horizontal axis). If the fiber length is 35 mm, it's projection may be 30 mm or less to the degree of fiber straightening in the fiber assembly (silver, lap, or yarn). Usually, the negative air pressure ranges from 65 mbar to 80 mbar.

Bibliography

Bracker, Technical spinning information, Private communication, Alexandria, Egypt.

Lawrence C.A., 2010, *Advances in Yarn Spinning Technology*, Woodhead Publishing, Cambridge, UK.

References

Elhawary I.A., 2010, Technology of rotor spinning, 2010 Lecture notes, Alexandria University, Alexandria, Egypt.

Elhawary I.A., 2014, Dynamic balancing of textile rotating masses, Post-graduate course, TED, Alexandria University, Alexandria, Egypt.

Elhawary I.A., 2015, The impact of the spinning technologies on the EUG-YQF, 2015 Belt Wide Cotton Conference (BWCC), San Antonio, TX.

El Moghazy Y. and Chewning C., Jr., 2001, *Cotton Fiber to Yarn Manufacturing Technology*, Cotton Incorporated, Cary, NC.

Lord P.R., 2012, *Handbook of Yarn Production*, Woodhead Publishing, Cambridge, UK.

McCreight D.J., Feil R.W., Booter Baugh J.H., and Backe E.E., 1997, *Short Staple Yarn Manufacturing*, Carolina Academic Press, Durham, NC.

Simpson W.S. and Crawshaw G.H., 2002, *Wool Science & Technology*, Woodhead Publishing, Cambridge, UK.

SKF Group, 2013, Roller Bearings Catalogue, Sweden.

Textilmaschinen-Komponenten, 1991, SKF Almanac, 7th revised edition, SKF Textilmaschinen- Komponenten, Stuttgart, Germany.

Appendix II: Constructional Aspects of Different Rotor Sets and Their Resonances Curves

Figures A2.1 through A2.3 show traditional and non-traditional rotor sets, similar to or the same as the rotor sets explained in Chapter 1. In Figure A2.2, the outer gland that surrounds the rotor spindle with its bearings is at a distance of $L_2 = 31.5$ mm from the rotor center of gravity S_R. It also has an elastic ring to minimize the noise level of the rotor set and to decrease the transmission of the rotor vibration to the rotor spinning machine's chassis and consequently to the floor of the spinning mill.

Figure A2.3 shows a non-traditional rotor that contains more than one elastic element that is located near to the upper bearing of the rotor's spindle. The effect of the elastic elements has been explained previously.

Figure A2.4 shows a three-peaked resonance curve that shows the relationship between the amplitude of vibrations of the rotor for a traditional rotor set in millimeters and its radian speed in rad/s. The three peaks can be explained as follows: the first peak (lowest critical speed) is for the wharve, the second peak (medium critical speed) is due to the rotor (cup/pot/camera), and the highest critical speed is due to the rotor's spindle as a short shaft with its mass distributed uniformly around its axis of rotation.

Refer to Chapter 1 for a review of these details.

Summary Points

1. The traditional rotor set consists of
   - A spindle supported by two ball bearings, one near the cup and the other near the wharve.
   - A rotor fixed to the spindle at one of its ends.
   - A wharve fixed to the other side of the spindle that drives the rotor with the spindle via a tangential belt drive.
   - A gland that surrounds the spindle with the bearings.
2. The free-body diagram of the spindle is considered to be an extended end beam (overhanging beam).
3. The non-traditional rotor set has a spindle without a wharve where the tangential belt drive meets the spindle directly.

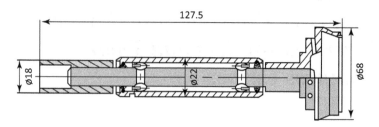

FIGURE A2.1
Traditional rotor set. (From Makarov A.I., 1968, *Construction & Design of Machines Used in Yarn Production*, Machines Design Press, Moscow, RFU.)

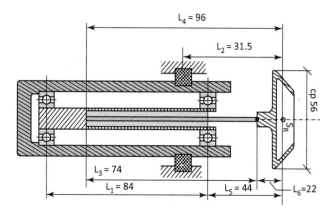

FIGURE A2.2
Non-traditional rotor set with an external elastic ring around the gland. (From Karitiski et al., 1974, *Vibration & Noise in Textile & Light Industry Machines*, Light Industry Press, Moscow, RFU.)

4. The new rotor set design implements elastic elements to reduce vibrations and noise levels, as well as to ensure smooth running at high speeds.

5. The rotor set resonance curve is a three-humped camel due to the three elements that form the rotor set, wharves, and spindle.

6. The bearings of the spindle and their outer and inner races are channeled in the body of the gland and spindle, respectively.

7. The traditional rotor set is self-lubricated by grease that is sufficient until the bearings are damaged or the rotor span life cycle finishes.

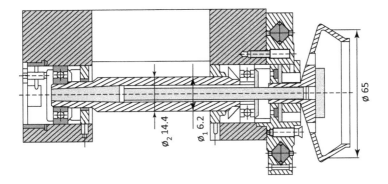

FIGURE A2.3
Non-traditional rotor set with elastic elements near the upper ball bearing. (From Karitiski et al., 1974, *Vibration & Noise in Textile & Light Industry Machines*, Light Industry Press, Moscow, RFU.)

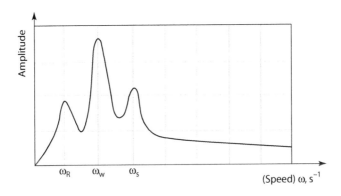

FIGURE A2.4
Qualitative resonance curve of a traditional ω_w (wharve critical speed), ω_R (rotor critical speed), and ω_s (spindle critical speed). (From Moscow State University of Design & Technology, Moscow, RFU, 1973.)

Review Questions

   Q#1: Using a line diagram, show the construction of a traditional rotor set.

   Q#2: Where are the elastic elements implemented in the rotor set?

   Q#3: Sketch the resonance curve of the rotor set?

Answers to Review Questions

Q#1: See Figure A2.1.

Q#2: They are usually implemented near the cup bearing where a high dynamic reaction is created (large rotor dynamic unbalance).

Q#3: See Figure A2.4.

Bibliography

Karitiski Ya.I., 1973, *Vibrations of Textile Machines*, Machines Design Press, Moscow, RFU.

References

Karitiski Ya.I., Kornev I.V., Lagynov L.Fy., Cyshkova R.I., and Khodekh M.I., 1974, *Vibration & Noise in Textile & Light Industry Machines*, Light Industry Press, Moscow, RFU.

Makarov A.I., 1968, *Construction & Design of Machines Used in Yarn Production*, Machines Design Press, Moscow, RFU.

Moscow State University of Design & Technology, Moscow, RFU, 1974.

Appendix III: An Experimental Determination of the Mass Moments of Inertia of the Rotor Spinning Machine's Rotor

A. Polar Mass Moment of Inertia

The polar mass moment of inertia can be determined using a torsional pendulum (Bevan, 1969) as shown in Figure A3.1.

The rotor is suspended with its axis vertical to a ceiling by three long flexible parallel inextensible wires or cords of equal length L. The cords or wires are fixed at the corners of equilateral triangles that are each at a distance a from the axis of the rotor when the rotor is torsionized about its vertical axis via a small angle θ and then released. The rotor will fluctuate (oscillate) with frequency f that is expressed through the system parameters or dimensions. As the cords or wires are suspended (fixed) in symmetry with the rotor axis or its center of mass, the tension in each cord will be 1/3 of the rotor's weight.

For an angular small displacement θ of the rotor, the angular displacement of each wire from the vertical direction is φ.

$$\ell \cdot \varphi \cong a \cdot \theta$$

that is,

$$\varphi = \frac{a}{\ell} \cdot \theta \qquad (A3.1)$$

The horizontal component of each wire tension is

$$\frac{w}{3} \cdot \tan \varphi \qquad (A3.2)$$

The horizontal component of wire tension is in equilibrium, that is, there is no resultant force.

These forces generate restoring torque T with a value of

$$3 \times \frac{w}{3} \cdot \tan \varphi \cdot a \cos\left(\frac{\theta}{2}\right) = w \cdot a \cdot \tan \varphi \cdot \cos\left(\frac{\theta}{2}\right) \qquad (A3.3)$$

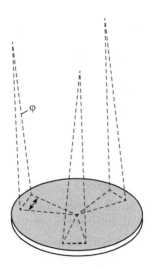

FIGURE A3.1
Torsional pendulum for determining I_0 of the rotor. (From Bevan T., 1969, *The Theory of Machines*, Longmans Press, London, UK and Sang AM Press, Calcutta, India.)

For small angles, $\tan \varphi = \varphi$ and $\cos(\theta / 2) = 1$.

The total restoring torque T is $w \cdot a \cdot \varphi \cdot 1$.

$\therefore$ By substituting φ from Equation A3.1:

$$T = W \cdot a \left(\frac{a}{\ell} \cdot \theta \right)$$

$$T = w \cdot \frac{a^2}{\ell} \cdot \theta \qquad\qquad (A3.4)$$

The restoring torque is directly proportional to the small angular displacement, that is, the motion is a simple harmonic motion (SHM); then, the torque T is

$$T = I \cdot \alpha \qquad\qquad (A3.5)$$

where:
   I = Mass moment of inertia of the rotor
   α = Angular accumulate of the rotor

$$\therefore \alpha = T/I = w \cdot \frac{a^2}{\ell} \cdot \theta / m_r \cdot K^2$$

$$= w \cdot \frac{a^2}{\ell} \cdot \theta \bigg/ \frac{w}{g} \cdot \leq k^2$$

$$= \frac{g}{\ell} \cdot \left(\frac{a}{k}\right)^2 \cdot \theta \qquad\qquad\text{(A3.6)}$$

where:
    k = Radius of gyration of the rotor mass
    g = Gravity accelerations

The frequency f of oscillations (fluctuations) is

$$f = \frac{1}{2\pi}\sqrt{\frac{\alpha}{\theta}}$$

$$= \frac{1}{2\pi}\frac{a}{k}\cdot\sqrt{\frac{g}{\ell}} \quad \text{Hertz (Hz)}$$

or

$$k = \frac{a}{2\pi f}\sqrt{\frac{g}{\ell}} \qquad\qquad\text{(A3.7)}$$

where:
    a = Radius of the rotor
    ℓ = Length of the cord (wire)

It is important to note the following:

a. The mass moment of inertia of the rotor about its axis of rotation is named the polar mass moment of inertia (I_0) and it can easily be determined experimentally using a torsional pendulum.

b. The mass moment of inertia around an axis perpendicular to the axis of rotation of the rotor is named the equatorial mass moment of inertia (I_q), but if it is passing through the rotor's mass center, then it is named the central equatorial mass moment of inertia of the rotor (I_{qc}).

Active example 1
A rotor of a rotor spinning machine has the following properties:

- Rotor mass = 70 g = 0.07 kg
- Rotor radius = 23 mm = 0.023 m
- Through experimental work using a torsional pendulum, it was found that 50 cycles took 83.2 s (seconds)
- Length of the cord = 3.0 m

Calculate the rotor's polar mass moment of inertia.

Solution

$$I_0 = m_R k^2 = 0.07 \times k^2$$

$$k = \frac{a}{2\pi f} \cdot \sqrt{\frac{g}{\ell}}$$

$$= \frac{0.023}{2\pi f} \cdot \sqrt{\frac{9.8}{3}}$$

$$= \frac{3.6624 \times 10^{-3}}{f} \times 1.8074$$

Time of cycle = 83.2/50 = 1.664 s/cycle.

$$f = \frac{1}{time} = 0.60096$$

By substituting $f = 0.60$ in the previous equation,

$$\therefore k = 0.011, \ k^2 = 1.217 \times 10^{-4}$$

$$\therefore I_0 = 0.07 \times 1.2171 \times 10^{-4}$$

$$= 8.52 \times 10^{-6} \ kg/m^2$$

B. Equatorial Mass Moment of Inertia

The equatorial mass moment of inertia can be determined experimentally using a torsional pendulum with two wires (cords) instead of cords. In addition, the determination of the center of gravity of the rotor is necessary. Practically, this is too difficult. Therefore, it is determined by considering the rotor as a thin disc, where we theoretically find

$$I_q = I_{qc} = \tfrac{1}{2} I_0 \tag{A3.8}$$

where I_q and I_{qc} are the equatorial mass moment of inertia of the rotor around the midpoint between the two wires or around the mass center of the rotor.

Active example 2

Calculate the equatorial (central equatorial) mass moment of inertia of a rotor of a rotor spinning machine using the data from Active Example 1.

Solution

$$I_q = I_{qc} = \tfrac{1}{2} I_0$$

$$= \tfrac{1}{2} \times 8.52 \times 10^{-6}$$

$$= 4.26 \times 10^{-6} \, \text{kg/m}^2$$

Summary Points

1. The massive geometrical characteristics of the rotor spinning machine's rotor are very important when studying rotor vibrations and dynamics. These are mass, center of gravity, and mass moments of inertia.

2. The rotor is considered a rigid body but its shaft (spindle) is an elastic body.

3. The rotor has two mass moments of inertia (Bedford & Fowler, 2008): (1) a polar mass moment of inertia around the axis of rotation or center of gravity; and (2) an equatorial mass moment of inertia around an axis perpendicular to the axis of rotation.

4. The radius of gyration is the constant of proportionality between the rotor's mass and the rotor's inertia (I_0), where $I_0 = m_R \cdot k^2$, $m_R =$ rotor's mass, $k =$ radius of gyration, and $I =$ inertia of rotor's mass (polar).

5. The frequency of oscillation (speed of oscillation) $= 1/T$ where T is the time of one cycle. Also, $T = 1/f$.

6. A rotor's balancing requires spindle removal.

Review Questions

Q#1: Define each of the following:
   a. Polar mass moment of inertia
   b. Center of gravity (mass center of the rotor)

Q#2: What is meant by the massive geometrical characteristics of the rotor?

Q#3: Explain an experiment for determining the mass moment of inertia of the rotor around its axis of rotation.

Q#4: How do you determine the equatorial mass moment of inertia of the rotor of the rotor spinning machine? (Dobrigy Roskova [ed.], 1958)

Answers to Review Questions

Q#1:

a. It is the rotor's mass inertia around its axis of rotation. It can be calculated theoretically or determined experimentally.

b. The center of gravity or rotor mass center is a point located in the rotor at which the rotor will be in balance horizontally when it is put on a prism. The load on the rotor (cup) is concentrated in that center as an inertia force (centrifugal force and gyroscopic moment).

Q#2: The massive geometrical characteristics of the rotor refer to all the items related to its mass and dimensions: mass moments of inertia, mass, and weight. All items concerning the dimension of the rotor are diameter, width, location of the center of gravity, and rotor eccentricity.

Q#3: See Figure A3.1 and its associated text.

Q#4: The equatorial mass moment of inertia is calculated by considering the rotor a thin disc and knowing the polar mass moment of inertia I_0 (experimentally).

Review Problems

Q#1: A rotor of a rotor spinning machine has a mass of 75 g and a radius of gyration $k = 0.014$ m. What is the value of I_0 and I_q (I_{qc})?

Q#2: The experiment using the torsional pendulum for a rotor gave a total time of 95 s for 60 oscillations. What is the value of frequency of oscillation?

Q#3: Write an equation for calculating the radius of gyration of the rotor.

Q#4: Derive an equation for calculating the oscillation frequency of a rotor subjected to an experiment with the torsional pendulum.

Answers to Review Problems

Q#1: $I_0 = mk^2 = 0.075 \times 0.014^2 = 1.47 \times (E-5) \, \text{kg/m}^2$

$$I_q(I_{qc}) = \tfrac{1}{2} I_0 = 7.35 \times (E-6) \ \text{kg/m}^2$$

It is assumed that the rotor is a thin disc.

Q#2: Time of 1 oscillation $t = 95/60 = 1.58$ s/cycle

The oscillation frequency $f = 1/T = 0.63$ oscillations/s

Q#3: $k = \dfrac{a}{2\pi f} \sqrt{g/l} \, \text{m}$

Q#4: $f = \dfrac{1}{2\pi} \cdot \dfrac{a}{k} \sqrt{g/l} \ \text{Hz}$

References

Bedford A.M. and Fowler W., 2008, *Engineering Mechanics: Dynamics*, 5th edition, Pearson Prentice Hall, Upper Saddle River, NJ.

Bevan T., 1969, *The Theory of Machines*, Longmans Press, London, UK and Sang AM Press, Calcutta, India.

Dobrigy Roskova C.O. (ed.), 1958, *Instruments & Stands for Research Work in Textile Machines*, Machines Design Press, Moscow, RFU.

Appendix IV: Dynamic Balancing of Rotors

SKF Technique

The lifespan and working conditions of a rotor depend to a large extent on an operation that is as free from mechanical vibration as possible. Such low-vibration operation is obtained by lowering the rotating rotor's unbalance to an allowable residual magnitude, the so-called "allowable residual unbalance" or, more simply, "residual unbalance."

The unbalance is characterized by the distribution of the rotor's material around its axis of rotation, that is, the rotational axis and the adjoining principle axis of inertia of the rotor do not coincide due to the rotor body's material distribution. Under working conditions, this lack of symmetry generates centrifugal forces, or inertia forces, that instigate different disturbing dynamic reactions in the rotor's bearings. Accordingly, immediate attention must be paid to increased noise and mechanical vibrations. The rotor's bearing is where excessive unbalance always leads to a much reduced lifespan of the bearings, as compared to when the unbalance of the rotor is maintained at a level of "residual unbalance."

In the case of the rotating mass shown in Figure A4.1, there is only one extra mass with its center of gravity s_1 at a distance r from the axis of rotation.

The principle axis of inertia (PAI) h shifts away toward the extra mass m (the shift = e from the rotational axis). When the body rotates with speed n in revolutions per minute (RPM), the extra mass m results in the following inertia force F:

$$F = m \cdot \omega^2 \cdot r \qquad (A4.1)$$

where:
- m = Value of extra mass in kilograms
- r = Extra mass center of gravity S, shift from axis of rotation c in meters, that is, there is a gap (distance between axis of rotation and mass centers)
- ω = Radian speed in radian/s
- F = Inertia force (centrifugal force) in N (newton)

Formula A4.1 shows the rapid increase of the inertia force as the square of the body's rotating speed.

The value of the centrifugal force is determined by the product of the extra mass m and the distance r from its center of gravity S from the axis of

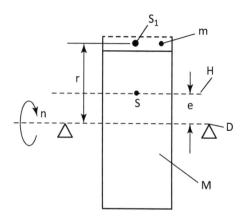

FIGURE A4.1
A rotating mass with uneven distribution around the axis of rotation. (From SKF Textilmaschinen-Komponenten, 1991, *SKF Almanac*, 7th revised edition, SKF Textilmaschinen-Komponenten, Stuttgart, Germany.)

rotation, multiplied by the square of the radian speed of the rotor (rotating body). The product $m \cdot r$ is called unbalance, that is,

$$U = m \cdot r \tag{A4.2}$$

where:
   U = Unbalance or static u unbalance in grams per millimeter
   m = Extra mass in grams (g)
   r = Actual (effective) radius in millimeters

For technological and other reasons, the pot of the rotor spinning machine is never manufactured with zero extra mass, as it is impossible to balance 100%. The rotor's mass distribution is not uniformly distributed around its axis of rotation. Balancing includes the determination of the size (value) and position (distance r) of the extra mass where it is possible to remove it (by means of drilling, milling, etc.) or add additional material to reduce it to the size (value) of the allowable residual unbalance. These trials or techniques could lead to a shift in the PAI to coincide more or less with the adjacent rotational axis.

The necessary standard of balancing depends on the type of rotor and its rotational speed. Rotor unbalancing can be carried out rapidly and reliably with an unbalance measuring machine.

SKF introduced to the spinning industry unbalance measuring machines for standard rotors for calibration.

IRD Mechanalysis

This technique is applied for the balancing of overhung rotors as rotors of rotor spinning machines. This type of rotor has its balance correction planes located outside the carrying bearings as shown in Figure A4.2. This rotor configuration often makes balancing difficult when one correction plane is applied at one time.

Generally, the rotor of a rotor spinning machine has a length to diameter ratio $\ell/d < 0.5$. This provides an opportunity for rotor balancing to be achieved via static unbalancing if its speed is less than 1000 RPM. The open-end spinning rotor has a running speed of over 1000 RPM, that is, up to 150 kRPM.

Therefore, it must be balanced in two planes. The IRD mechanalysis technique is shown in Figure A4.2.

Balancing Outboard Rotors

The balancing procedure is as follows:

1. Stop the rotor.
2. Fix the reference phase that is glued to cardboard on bearing 1.
3. Make a reference mark on the new selected plane A.
4. Fix the two pickups of the IRD 350 vibration on the bearing housing 1 and 2.
5. Run the rotor to its working speed and record the original unbalance.
6. Stop the rotor.

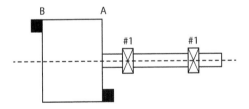

FIGURE A4.2
The outboard rotor configuration of a rotor spinning machine. (From IRD Mechanalysis Instruction Manual, md@ indmech. of Mumbai, India.)

7. Add an atrial weight to the near selected plane at a known radius and location.

8. Record the new readings of the value and location of the correct weight and add it to plane A.

9. If the amplitude at 2 indicates that two-plane balancing is required (either by single-plane balancing or the two-plane balancing technique), add a couple of masses at 180° to the planes A and B to maintain the balance at the plane without any interference.

10. The balancing procedure can be repeated two or more times to diminish the amplitude of vibration of the bearings' housing.

The rotor balancing quality can be checked industrially or practically through IRD Mechanalysis information such as the unbalance tolerance guide for rigid rotors based on the Association of German Engineers (VDI) standards where it has grades from G 0.4 (ultra-precision balancing) to G 40 (passenger car, wheels, and rims).

In addition, as shown in Broch (1980), the balancing quality can be checked practically using the maximum residual unbalance chart as laid down in ISO 1940 (1973).

Summary Points

1. The unbalance of the rotor is due to the uneven distribution of its mass around its axis of rotation.

2. The centrifugal force F (in N) due to unbalance is $F = mr \cdot w^2$, where mr = static unbalance in grams per millimeter, r = distance of extra mass or overmass from the axis of rotation of the rotor, and w = rotor radius speed.

3. The principal axis of inertia (PAI) is the axis around which the rotor mass is distributed. If the distribution is even, then the PAI coincides with the axis of the rotor's rotation, that is, theoretically, the rotor is dynamically balanced precisely with residual unbalance. In general, the PAI tends to stay near the overmass (extra mass).

4. The rotor's unbalance generates a dynamic reaction in the rotor spindle's ball or roller bearings that reduces their lifespan due to wear and heat effects.

5. The rotor's unbalance also creates a forced vibration associated with noise that will disturb and harm the working conditions. The workers and other staff will also suffer.

6. Standard balancing depends on the massive geometric characteristics of the rotor and its running speed.

7. After any balancing process, it is important to never have a zero value of residual unbalance.

8. The quality of balancing is affected by the value of residual unbalance. As it is minimized, the balancing quality will increase, but the cost is directly proportional to the balancing quality.

9. Some international organizations have introduced unbalance measuring machines to the rotor spinning industry, such as IRD, B&K, SKF, and so on.

10. The IRD Mechanalysis technique for balancing outboard rotors could be applied for the rotor of the rotor spinning machine.

11. The in-place balancing suggested by IRD Mechanalysis could be used with certain precautions for rotor balancing during its operation in a rotor spinning machine to save time and increase the balancing quality.

12. The IRD Mechanalysis technique measures the forced vibration at the rotor spindle bearings' housing using several trials, and then trial masses can be added.

13. The accumulation (building up) of trashes in the groove of the rotor affects the rotor unbalance. In addition, the peeling-off point of the trashes has another effect.

Review Questions

Q#1: What is meant by the following: overmass, static unbalance, residual unbalance, balancing quality, and outboard rotors?

Q#2: Explain the concept of IRD Mechanalysis for outboard rotor (as rotors of open-end spinning machines) balancing.

Q#3: Write the steps for rotor balancing using the IRD Mechanalysis technique for outboard rotor balancing.

Q#4: Write the equation of the centrifugal force exerted on a rotor with overmass (extra mass).

Q#5: Mention the drawbacks of rotor unbalance.

Answers to Review Questions

Q#1:

- Overmass (heavy spot) is a mass that leads to the uneven distribution of total mass.
- For static unbalance, see Equation A4.2.

Q#2: The concept depends on measuring the vibrations level at the bearing housing and using trial masses.

Q#3: See Figure A4.2 with the associated steps.

Q#4: See Summary Point 2.

Q#5: The drawbacks are

   a.   The increase in the level of the forced vibration
   b.   The increase of the dynamic reactions in the rotor bearings
   c.   The damage of healthy bearings
   d.   The increase of the noise level and its associated drawbacks

Review Problems

Q#1: Write about the relationship between the rotor rotational speed in RPM and its speed in radian per second.

Q#2: If a rotor mass is 62 g with a radius equal to 24 mm, a rotational speed of 30 kRPM, and the rotor mass center of gravity has shifted from the axis of rotation by a distance of 60 µm (micrometer), what will be the centrifugal force on the rotor mass?

Q#3: In Problem 2, assume an overmass of 2 g on the rotor collecting surface (trash buildup). What is the value of the inertia force (centrifugal force)?

Q#4: In Problems 2 and 3, calculate the value of static unbalance u = me (mr) in g/m.

Q#5: A rotor with a mass =38 g has a centrifugal force = 48 N. What is the value of that force if the mass is increased to 67 g?

Answers to Review Problems

Q#1: If the rotor's rotational speed = n RPM, then its radius speed (ω rad/s) will be

$$\omega = n \times \frac{2\pi}{60} = n/9.55$$

that is,

$$n = 9.55 \times \omega$$

$$\cong 10.\omega$$

Q#2: $\omega = (30 \times 1000) \times \dfrac{2\pi}{60} = 3140 \text{ rad/s}$

or

$$\omega = \frac{n}{10} \cong 30 \times 1000/10 = 3000 \ (\text{error} = 5\%)$$

$$m = 0.062 \text{ kg}$$

Shift = eccentricity = $e = 0.062 \times 10^{-3}$ m

$$\therefore cF = 0.062 \times (3140)^2 \times 0.062 \times 10^{-3}$$

$$= 37.9\,N\,(\cong 38\,N)$$

Approximately, we can assume $\omega = 3000$ rad/s.

Then, cF = 35 N (error = 9%).

It is important to note the following:

- One micrometer (μm) = 10^{-6} m
- cF = Centrifugal force

Q#3: The trash distance, that is, the rotor radius = 24 mm (0.024 m)

$$cF = 0.002 \times 3140^2 \times 0.024$$

$$= 473\ N$$

The large value of cF is due to the trashes that are built-up on the collecting surface of the rotor (r = 24 mm).

Q#4: $u = m \times e = 62 \times 0.06 = 3.72$ g/mm

$$u = m \times r = 2 \times 24 = 48 \text{ g/mm}$$

$$cF \text{ for } 78\ g = 48 \times \frac{67}{38} = 84.6\,N \cong 85\,N$$

Q#5: cF = 85 N for 78 g.

References

Broch J.T., 1980, *Mechanical Vibration and Shocks Measurements*, Brüel and Kjær, Naerum, Denmark.

IRD Mechanalysis Instruction Manual. md@ indmech.com, Mumbai, India.

SKF Textilmaschinen-Komponenten, 1991, *SKF Almanac*, 7th revised edition, SKF Textilmaschinen-Komponenten, Stuttgart, Germany.

Appendix V: Noise Level of the Rotor Spinning Machine

IRD Mechanalysis has discussed several topics concerning noise. Noise can be psychologically defined as an "undesirable sound." This would include a drop from a faucet at midnight or the roar of a rocket as it leaves the launch pad.

But technically, noise or sound is a pressure fluctuation in the air that radiates far away from the source.

There are several causes of noise in a rotor, such as the vibration of the solid constructions of the rotor spinning machine; thus, mechanical vibrations in machinery are a source of noise generation. The turbulent air flow when mixed with low air flow can generate noise or undesirable sounds. These sources of noise are commonly known in the spinning industry and knowledge of them can make it possible to minimize the noise to an acceptable level.

Noise characteristics can be described by examining air pressure fluctuations. The first characteristic is the speed propagation "c," which is the speed of soundwaves that radiate. The noise waves' speed of radiation is equal to 346 m/s at a standard temperature and pressure and it is independent from the sound amplitude.

Another characteristic of sound is the frequency "f." It can be defined as the number of noise waves in a specified period of time in seconds or minutes.

The frequency of noise is measured in relation to human hearing and it is expressed in cycles per second (CPS) or hertz according to an international standard. With respect to human hearing, sound frequencies that range from Hz to 20 kHz are named audio-sonic, which is concerned with industrial noise and has the greatest effect on humans.

The third noise characteristic is the wavelength. The relationship between the wavelength h, frequency f, and propagation speed c is

$$f = c / \lambda \qquad\qquad (A5.1)$$

Thus, the sound frequency is inversely proportional to wavelength. Very high frequency noises have a short wavelength, and vice versa.

The noise frequency is important in identifying a noise source. The most familiar reason for noise is mechanical vibrations in the different parts of a rotor spinning machine. The frequency at which these parts vibrate determines the frequency of the noise radiated, for example, a 50 Hz (3000 CPM) vibration will generate a 50 Hz (3000 CPM) sound wave.

When a noise is generated in the absence of mechanical vibrations in the rotor spinning machine, it is referred to as "inherent" to machine operation.

If this is actually the case, then the noise level must be controlled by other techniques such as barriers to noise and acoustic enclosures that absorb the noise. Lastly, the wavelength is a vital factor to consider for microphone location when measurements are processed.

The measure of the noise amplitude is a measure of how far the air pressure is above (over) the atmospheric pressure and then drops below the atmospheric pressure. The deviation of the sound pressure over and under the atmospheric pressure is known as amplitude, that is, the sound amplitude is a measure of the air pressure over and under the atmospheric pressure. The noise pressure is directly proportional to the noise amplitude. The sound pressure amplitudes are expressed in "micro atmospheres," that is, micro atmospheric pressure (1 bar = 1 atm). Also, it can be expressed in dynes per square centimeter.

The minimum sound pressure for human hearing is 0.0002 microbars (threshold of hearing), while the maximum air pressure that the human ear can tolerate is $\cong$2000 microbars. This is named the "threshold of pain."

The decibel (dB) is a simple measure for sound pressure amplitude where a logarithmic decibel scale is used. The sound pressure level (SPL) or dB is defined by

$$SPL(dB) = 20 \log_{10} \frac{\text{Sound pressure}}{\text{Std. ref pressure}} \tag{A5.2}$$

where the standard reference pressure is the minimum air pressure for human hearing, that is, "the acute threshold" = 0.0002 microbars. The measured sound pressure is the sound amplitude of interest. From the previous equation, the acute threshold is 0 dB while the threshold of pain is 140 dB. This scale ranges from 0 to 140 dB based on the tolerability of noise levels. There are different charts for common non-industrial noise levels and for industrial applications. Another interesting topic in noise levels is the use of equipment that measures sound in a similar code of human hearing. Human ears do not hear equally well at standard frequencies and amplitudes. Noise measuring apparatuses that approximate the response of the human ear with the aid of electronic "weighting networks" to familiar weighting networks embodied in sound evaluation equipment include the "A," "B," and "C" weighting networks. The response characteristics of these three weighting networks are listed in a special chart (IRD Mechanalysis). The "A" weighting network is the closest to the human ear with low sound levels that are less than 55 dB; the "B" weighting network contains medium sound levels that range from 55 to 85 dB; and the "C" weighting network contains sound levels that are over 85 dB.

In general, the sound level (noise level) readings establish compliance with noise control legislation with the use of the "A" weighting network, irrespective of the sound level. Also, hearing damage (loss) is highly correlated with weighting network equipment. The "C" weighting is generally used where there is a requirement for relatively low frequency noise. The "B" weighting is not used except in special circumstances. For further reading, see IRD Mechanalysis and Broch (1980). More interesting studies concerning noise level can be found in the works of Joëlle Courrech (Broch, 1980) and Karitiski et al. (1974).

Noise Levels in dB of Different Mechanisms of a Rotor Spinning Machine

The following noise levels are given by Karitiski et al. (1974):

1. **Rotor and its drive:**
   - General noise level: 9–90 dB
   - Octave level of frequencies in hertz:
     63 (64–67), 125 (70–75), 250 (75–80), 500 (82–82), 1000 (84–86), 2000 (81–84), 4000 (78–81), and 8000 (77–78)

2. **Motor of the opening cylinder:**
   - General noise level: 81–87
   - Octave level of frequencies in hertz:
     63 (53–64), 125 (63–72), 250 (68–81), 500 (72–81), 1000 (76–83), 2000 (75–78), 4000 (70–75), and 8000 (69–73)

3. **Winding head shaft:**
   - General noise level: 81–85
   - Octave level of frequencies in hertz:
     63 (51–56), 125 (54–55), 250 (59–72), 500 (67–70), 1000 (76–76) 2000 (74–74), 4000 (66–78), and 8000 (64–71)

4. **Feeding rolls shaft:**
   - General noise level: 72–84
   - Octave level of frequencies in hertz:
     63 (47–61), 125 (57–64), 250 (63–69), 500 (56–76), 1000 (67–76), 2000 (65–74), 4000 (61–75), and 8000 (71–76)

5. **Overall rotor spinning machine noise levels:**
   - General level: 90–91
   - Octave level of frequencies in hertz:
     63 (63–68), 125 (70–74), 250 (75–83), 500 (82–83), 1000 (86–86), 2000 (82–85), 4000 (80–82), and 8000 (79–80)

General Notes

IRD Mechanalysis introduced to the industry two typical standards: Figure A5.1 gives industrial noise levels while Table A5.1 gives the maximum allowable exposure time to the noise.

Summary Points

1. Psychologically, noise is defined as an unwanted (undesirable) sound.

2. Mechanical vibrations are a main reason for the creation or generation of industrial noise.

3. One of the techniques used to minimize noise is to establish a program for controlling mechanical vibrations. Also, incorporating elastic elements into the machine's movable parts can help in lowering the noise level.

4. Inherent noise is that which is generated when the machine is highly controlled with regard to mechanical vibrations. Noise control

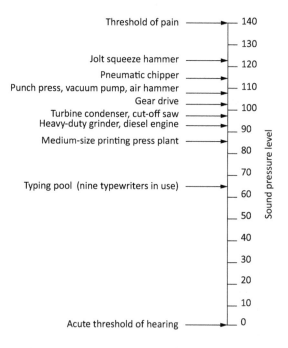

FIGURE A5.1

Typical and industrial noise levels. (From IRD Mechanalysis, 1985, Audio-visual training, Special instruction manual, md@ indmech.com, Mumbai, India.)

TABLE A5.1

Typical Exposure Limits of Noise

Permissible Exposures	
Duration Per Day in Hours	Sound Level dBA
8	90
8	92
4	95
3	97
2	100
1½	102
1	105
½	110
¼ or less	115

Source: IRD Mechanalysis, 1985, Audio-visual training, Special instruction manual, md@ indmech.com, Mumbai, India.

here involves the use of noise barriers or caustic enclosures that absorb sound.

One micro = 10^{-6}, one bar = 10^{-5} Pa (Pa = N/m²).

5. The measure of noise amplitude is a measure of how far the air pressure is above the atmospheric pressure (one bar = 10^{-5} Pa (N/m²).

a. Noise pressure is directly proportional to noise amplitude.

b. The threshold of pain in noise is 140 dB.

c. The acute threshold is 0 dB, that is, the minimum air pressure that the human ear can withstand.

6. The weighting network instruments for measuring noise are referred to as the "A," "B," and "C" weighting networks. The most applicable weighting network apparatus is "A."

7. There are two familiar maintenance companies in vibration and noise control:

a. IRD Mechanalysis (Ohio, USA)

b. B&K (Naerum, Denmark)

Review Questions

Q#1: What is meant by

a. Noise

b. The acute threshold of sound

c. The threshold of pain

Q#2: Explain the engineering characteristics of rotor spinning machine noise. What is the mathematical relationship between them?

Q#3: Summarize the different weighting networks used in noise measurement instruments.

Q#4: Specify the noise level ranges in decibels for each of the following mechanisms of the rotor spinning machine:

    a. Rotor and its drive

    b. Opening cylinder motors

    c. Feeding roll's shaft

    d. Overall rotor spinning machine noise

Q#5: What are the causes of noise in the rotor spinning machine?

Answers to Review Questions

Q#1:

    a. Psychologically, noise can be defined as an unwanted (undesirable) sound. The rotor spinning machine is a noise creator in the spinning mill.

    b. The acute threshold of sound refers to when the air pressure (amplitude) is 0 dB.

    c. The threshold of pain in noise is when the air pressure (amplitude) reaches 140 dB.

Q#2: An engineering noise characteristic of the rotor spinning machine is the speed propagation "c," which is the speed at which the sound wave radiates. This is independent from the sound amplitude. Another noise characteristic is the frequency "f," which is the number of noise waves in a specified time in CPS (hertz). The third noise characteristic is the wavelength "h."

The relationship between the three noise characteristics is shown in Equation A5.1.

Q#3: Noise measuring apparatuses are designed to approximate the responses of the human ear with the aid of electronic "weighting networks," which format the apparatuses' responses in a similar way to the human ear. The familiar "weighting networks" used in sound evaluation equipment are the "A," "B," and "C" weighting networks. The response properties of these weighting networks can be checked with the use of a special chart.

Q#4:

a. Rotor and its drive (31 kRPM):

    • The general noise level is 89–90 dB.

    • The highest noise level is at octave frequency 1000 (Hz) or 84–86 dB.

b. Combing roller's motor:
- The general noise level is 81–87 dB.
- The highest noise level is at octave frequency 1000 (Hz) or 76–83 dB.

c. Feeding roll shaft:
- The general noise level is 72–84 dB.
- The highest noise level is at octave frequency 8000 (Hz) or 71–76 dB.

d. Overall rotor spinning machine:
- The general noise level is 90–91 dB.
- The highest noise level is at octave frequency 8000 (Hz) or 86–86 dB.

Q#5: The sources of rotor spinning machine noise are

   The motor of the combing rolls with the tangential belt
- The rotor and its driving motor with the tangential belt
- The feeding roll shaft
- The winding head
- The suction fan that creates negative air pressure inside the machine

Review Problem

Q#1:

a. Write the equation for the sound pressure level (SPL) in decibel (dB).

b. Find the SPL in decibels at the acute threshold.

c. Calculate the SPL in decibels at the pain threshold.

Answers to Review Problem

Q#1:

a. The equation for the SPL in decibels is

$$\text{SPL} = (\text{dB}) = 20 \log_{10} \frac{\text{Sound pressure}}{\text{Std. ref pressure}}$$

   where std. reference pressure is the acute threshold = 0.0002 micro-bars ($=0.0002 \times 10^{-2} \times 10^5 = 2 \times 10^{-5}$).

   Always substitute std. ref pressure = 2×10^{-5} pa (N/m²)

   Sound pressure is the sound amplitude of interest.

b. For the acute threshold, the standard reference is

$$=\therefore \text{SPL}(\text{dB}) = 20 \ \log_{10}(1.0) = 20 \times \text{zero}$$

$$= 0, \text{ that is, } 0.0 \text{ dB}$$

c. The air pressure amplitude for the threshold of pain = 2000 microbars

$$(2000 \times 10^{-6} \times 10^5 = 200)$$

The ratio of pressure amplitude/acute threshold = $200/2 \times 10^{-5} = 10^{-7}$.
∴ The threshold of pain SPL (dB) = $20 \log 10^7 = 20 \times 7 = 140$ (140 dB)

References

Broch J.T., 1980, *Mechanical Vibrations & Shock Measurements*, Brüel and Kjær, Naerum.
IRD Mechanalysis, 1985, Audio-visual training, special instruction manual, md@indmech.com, Mumbai, India.
Karitiski Ya.I., Kornev I.V., Lagynov L.Fy., Cyshkova R.I., and Khodekh M.I., 1974, *Vibration & Noise in Textiles & Light Industries*, Light Industry Press, Moscow, RFU.

Appendix VI: Mechanical Vibrations

Introduction

The following definitions are important:

- The vibration is a periodic motion of an elastic body (rotor's spindle) when it is displaced from the equilibrium position.
- Free vibration is the periodic motion of the rotor's spindle (elastic body) when the movement is maintained by an elastic restoring force, that is, the vibration is free from external periodic loads applied to the elastic body.
- Forced vibration takes place due to the application of an external intermitted force (load).
- Damped vibration is that which passes/fizzles out with time as a result of the presence of external or internal frictional loads (forces).
- The undamped vibration can run continuously to infinity (∞) by ignoring the friction effects.
- When the vibrating mechanical systems have one degree of freedom, they need only a single coordinate (y) to explain the complete position of the system at any time (t).

Conservation of Energy

The free undamped mechanical vibration of a rotor-spindle set is due to the restoring forces (load) and forces of gravity of the rotor as a rigid body; as both types of forces are conservative, the principle of energy conservation is applied to obtain the rotor-spindle set's natural frequency. For small vibrations with a single-degree-of-freedom (as a conservative system) relation to its stable equilibrium (horizontal position), are determined by equation-free undamped mechanical vibration (Bedford & Fowler, 2008) :

$$\frac{d^2y}{d+2} + \omega^2 y = 0 \qquad (A6.1)$$

where ω as a constant depends on the rotor spindle's spring constant, can be determined by the characteristics of the rotor-spindle set, or is a circular natural frequency.

The general solution of differential Equation A6.1 is

$$y(t) = A \sin \omega t + B \cos \omega t \tag{A6.2}$$

Whether A and B are constant can be determined from the boundary condition of the set or system.

Another form of solution to Equation A6.2 is

$$y(t) = A_{\circ} \cdot \sin(\omega t - \theta) \tag{A6.3}$$

where $A_{\circ} = \sqrt{A^2 + B^2}$ is the *amplitude* of the rotor-spindle set vibration. The period of vibration is $\tau = \dfrac{2\pi}{\omega}$, which is the time it takes to finish one complete vibration or cycle; the frequency can be defined as the number of completed cycles per unit of time. It is the reciprocal of the period $f = \dfrac{1}{\tau} = \dfrac{\omega}{2\pi}$. In general, the frequency is expressed in cycles per second (CPS) or hertz (1 Hz = 1 CPS). Then, we can determine the characteristics of the mechanical vibrations of the rotor-spindle set: amplitude, period, and frequency of damped mechanical vibration.

The damping theorem states that frictional forces damp out the vibration. The damping effect will be embodied in Equation A6.1. Small mechanical vibrations of any damped single-degree-of-freedom system in relation to the stable condition are governed by the equation:

$$\frac{d^2y}{d+2} + 2g\frac{dy}{dt} + \omega^2 y = 0 \tag{A6.4a}$$

If $g < \omega$, then the rotor-spindle set is said to be subcritically damped. In such a case, the solution of Equation A6.4 is

$$y(t) = e^{-gt}\left(A \sin \omega_d t + B \cos \omega_d t\right)$$

where A and B are constants and ω_d is

$$\omega_d = \sqrt{\omega^2 - g^2} \tag{A6.4b}$$

The frequency and vibration period T are

$$f_d = \frac{\omega_d}{2\pi} \text{ and } \tau_d = \frac{2\pi}{\omega_a}$$

From Equation A6.4a, it is clear that $\omega_d < \omega$, so the period of oscillation is increased while its frequency is decreased as a result of subcritical damping.

Another type of damping (subcritical) when $g > \omega$ means that the vibrating system is considered to be subcritically damped. The general solution of Equation A6.4a in such a case is

$$y(t) = Ce^{-(g-h)^t} + De^{-(g+h).t}$$ (A6.4c)

where C and D are constants and h is

$$h = \sqrt{g^2 - \omega^2}$$ (A6.4d)

In the last case of critical damping, when $g = \omega$, then the general solution is

$$y(t) = Ce^{-dt} + Dte^{-dt}$$ (A6.5)

where C and D are constants.

The rate of damping is usually expressed in terms of a logarithmic decrement δ that is the natural logarithm of the amplitudes of damped vibrations at time t and time $t + \tau_d$.

$$\therefore \delta \ell n \frac{e^{-gt}}{e^{-g(t+\tau_d)}}$$ (A6.6)

The forced mechanical vibrations in this case mean that the rotor-spindle set is subjected to external forces (centrifugal forces on the rotor as a rigid body). The forced vibrations (mechanical) for a single-degree-of-freedom system are governed by the equation:

$$\frac{d^2y}{dt^2} + 2g\frac{dy}{dt} + \omega^2 y = a(t)$$ (A6.7)

where g and ω are given and (t) is a forcing function. The general solution of Equation A6.7 is composed of two parts—particular and homogeneous solutions:

$$\therefore y(t) = y_h(t) + y_p(t)$$

The particular solution $y_p(t)$ of Equation A6.7 is any solution that satisfies Equation A6.7. The homogeneous solution y_h is the general solution of Equation A6.7 when the right-hand side is set to zero.

━━━━━━━━━━━

The Disturbing Force Function

If we consider that the vibratory motion of the rotor set is subjected to a disturbing force function, then we can determine the response of the rotor set as a function of the disturbing force frequency.

━━━━━━━━━━━

The Particular Solution

In Equation A6.7, the sinusoidal function a_t is the disturbing force function that has the form $a(t) = a_0 \sin \omega_0 t + b_0 \cos \omega_0 t$.

Where a_0, b_0, and ω_0 (the frequency of the disturbing force) are considered constants, the particular solution is

$$y_{p(t)} = \left[\frac{\left(\omega^2 - \omega_0^2\right)a_0 + 2g\omega_0 b_0}{\left(\omega^2 - \omega_0^2\right)^2 + 4g^2\omega_0^2} \right] \sin \omega_0 t$$

$$+ \left[\frac{\left(\omega^2 - \omega_0^2\right)b_0 - 2g\omega_0 a_0}{\left(\omega^2 - \omega_0^2\right)^2 + 4g^2\omega_0^2} \right] \cos \omega_0 t$$

(A6.8)

And the corresponding amplitude is

$$A_p = \frac{\sqrt{a_0^2 + b_0^2}}{\sqrt{\left(\omega^2 - \omega_0^2\right)^2 + 4g^2\omega_0^2}}$$

(A6.9)

The following notes are interesting:

1. The resonant frequency is the frequency where the amplitude of the disturbing force is at its maximum.
2. The vibratory motion of the rotor set under the impact of the disturbing force (forcing function) is also called the steady-state solution. In addition, the motion (vibratory type) approaches the steady-state solution as time increases.

Summary Points

1. The differential equation of the motion of a vibration system (small undamped free vibration) always takes a well-known form as in Equation A6.1.
2. The different types of differential equations are
   a. Free undamped mechanical vibrations

$$\frac{d^2y}{dt^2} + \omega^2 y = 0$$

   b. Damped mechanical vibrations

$$\frac{d^2y}{dt^2} + 2g\frac{dy}{dt} + \omega^2 y = 0$$

   c. Forced damped mechanical vibrations

$$\frac{d^2y}{dt^2} + 2g\frac{dy}{dt} + \omega^2 x = a(t)$$

   d. Sinusoidal disturbing force
      The sinusoidal function $a(t)$ is

$$a(t) = a_0 \sin \omega_0 t + b_0 \cos \omega_0 t$$

   It is assumed that a_0, b_0, and ω_0 (disturbing force frequency) are constant.
3. The three types of damping are

   - Subcritical, when $g > \omega$, that is, $\omega_d = \sqrt{\omega^2 - g^2}$

   - Period $\tau_d = \dfrac{2\pi}{\omega_d}$ and $f_d = \dfrac{\omega_d}{2\pi}$

   - Subcritical damping, when $g > \omega$, that is, $h = \sqrt{g^2 - \omega^2}$
   - Critical damping, when $g = \omega$. The rate of damping is

$$\delta = \ell n \left[\frac{e^{-dt}}{e^{-d(t+\tau)}} = d\tau_d, \tau_d = \frac{2\pi}{\omega_d} \right]$$

4. The amplitude of forced damped vibration is

$$A_p = \frac{\sqrt{a_0^2 + b_0^2}}{\sqrt{\left(\omega^2 - \omega_0^2\right) + 4g^2\omega_0^2}}$$

The characteristics of the vibration are

- Amplitude value that depends on vibration type
- Period $\tau = \dfrac{2\pi}{\omega}$ second (s)
- Frequency $f = \dfrac{1}{\tau} = \dfrac{\omega}{2\pi}$ Hz (CPS)

5. The solution of the free vibration differential equation gives a simple harmonic motion (SHM).

6. The damping of vibration can be due to

   a. Internal friction in the material of the vibratory system.

   b. An external friction (a viscous-resistant medium) as dash pot.

7. The unbalance of the vibratory system creates forced mechanical vibrations.

8. The free vibration system helps in calculating the natural frequency that determines resonance.

9. The vibration of the rotor of the rotor spinning machine is a forced damped vibration.

10. American engineers say that all machines vibrate (forced damping).

11. The correlation is great between the amplitude of the forced vibrations (mechanical) and the noise amplitude pressure.

Review Questions

Q#1: Write the differential equation of free undamped vibration.

Q#2: Define the period, frequency, and amplitude of free undamped mechanical vibrations.

Q#3: What is meant by SHM?

Q#4: What is the resonance frequency?

Q#5: Differentiate between subcritical and critical damping.

Q#6 What is the logarithmic decrement?

Q#7: When do forced vibrations of a rotor set take place?

Answers to Review Questions

Q#1: See Equation A6.1.

Q#2: The period of vibration (mechanical) is

$$\tau = \frac{\omega}{2\pi}$$

The frequency of vibration is the number of cycles per second (CPS) or it is the reciprocal of τ, that is, $f = \frac{1}{\tau}$ the amplitude of the mechanical vibrations $A_0 = \sqrt{A^2 + B^2}$, and it is the displacement of the vibratory system from its stable equilibrium position to a certain value in y direction.

Q#3: SHM (simple harmonic motion) is the displacement time of the vibratory system in free undamped vibration due to the formula:

$$y(t) = A_0 \sin(\omega_t - \theta)$$

where A_0 is the amplitude.

Q#4: The resonance frequency is the frequency at which the displacement (amplitude) is at infinity or maximum.

Q#5: See Equations A6.4a–c and A6.5.

Q#6: See Equation A6.6.

Q#7: Due to the eccentricity (shift of the rotor center of gravity from its rotational axis with the rotational speed of the rotor set), the inertia force (centrifugal force) starts to create forced vibrations.

Review Problems

Q#1: The free undamped vibration differential equation is

$$\frac{d^2y}{dt^2} + \omega^2 = 0$$

Calculate:
a. The circular natural frequency
b. The linear natural frequency
c. When the critical speed of the system is attained
d. The period of oscillation

Q#2: Equation A6.2 has two constants, *A* and *B*. For initial conditions $y_0 = 0$ and $\dot{y}_0 = 0$, rewrite Equation A6.2 in its new form.

Q#3: Write the differential equation of the free damped vibration of a damped rotor set. Explain the different phases of damping of the rotor-spindle set.

Q#4: Write the equation for the damping rate.

Answers to Review Problems

Q#1:

a. Equation A6.1 could be written as

$$1 \times \frac{d^2y}{dx^2} + 1 \times y = 0$$

If the mass $m = 1$ (kg) and the spring constant $k = 1$ (N/m), then

$$\omega = \sqrt{\frac{k}{m}} = \sqrt{\frac{1}{1}} = 1.0 \, \text{rad/s} = 1.0 \, \text{s}$$

∴ The circular natural frequency $\omega = 1.0$ s

b. The linear natural frequency *f* is

$$f = \frac{\omega}{2\pi}$$

that is,

$$f = \frac{1}{2\pi}\sqrt{\frac{k}{m}}$$

$$\therefore f = \frac{1}{2\pi} = 0.159 \, \text{Hz (CPS)}$$

c. When the rotational speed of the rotor's spindle equals 1.0 s

d. The period of oscillation $\tau = f^{-1} = 2\pi$ s/cycle

Q#2:

$$y(0) = A \sin \omega \times 0 + B \cos \omega \times 0$$

$$= 0 + B \times 1$$

$$\therefore B = y(0)$$

$$y^{\circ}(t) = A\omega\cos\omega t \pm B\omega\sin\omega t$$

$$y^{\circ}(0) = A\omega \times 1 - B\omega \times 0$$

$$o = A\omega - 0$$

$$\therefore A = 0$$

The new form of Equation A6.2 is

$$y = y_0\cos\omega t$$

Q#3:

See Equations A6.4, A6.4a, A6.4b, A6.4c, and A6.5. Two phases of sub-critical damping are $g\langle\omega$ and $g\rangle\omega$.

Critical damping is when $g = \omega$. The mechanical vibration damping decreases the circular natural frequency ω to ω_d (Equation A6.4a).

Q#4: See Equation A6.6.

Bibliography

Broch J.T., 1980, *Mechanical Vibration and Shock Measurements*, Brüel and Kjær, Naerum, Denmark.

ENTEK IRD, International Corporation, 1996, *Introduction to Vibration Technology*, Mechanalysis Inc., Columbus, OH.

Seto W.W., 1964, *Schaum's Outline of Theory and Problems of Mechanical Vibrations*, McGraw-Hill Book Company, New York.

Reference

Bedford A.M. and Fowler W., 2008, *Engineering Mechanics: Dynamics*, 5th edition, Pearson Prentice Hall, Upper Saddle River, NJ.

Appendix VII: Lubrication

Lubrication involves using a lubricant between the rubbing surfaces of specified machine elements to reduce or prevent actual surface contact to minimize wear and consequently achieve a lower coefficient of friction.

Oils and greases are the most applicable lubricants despite the possibility of using any material that has the required viscosity properties. Usually, lubricants such as soap stones, graphite, or other greasy non-abrasive solids can be used. Even under some conditions, gases can create good lubricants.

The initial functions of lubricant are: to prevent metal-to-metal contact; to encourage uniform temperature conditions via the bearing by carrying away the generated heated locally; to resist the entrance of foreign impurities between the sliding surfaces; and to fight against the corrosion of highly worked surfaces. The selection of a lubricant involves choosing between an oil or a grease. The safe velocity limits up to which rolling elements can work depend on creating and maintaining oil films. A great variety of greases and oils will provide sufficient performance in practice except under extreme conditions. Oil is a better lubricant as it efficiently carries the generated heat away, can be fed readily between contact areas, and expels trash and foreign matter.

Oils are applied to high-velocity working elements where the temperature is greater than 1000°C. Despite this, the general direction is to use grease in bearings, as is the case in the rotor and combing roll bearings in the rotor spinning machine, due to their simplicity in housing design, better seal against pollution by dirt and moisture, ease of handling and replenishment, and ease of cleaning where the pollution of final products is not desirable. A certain merit of greases is the ease by which additives or inhibitors can be embodied. Extreme pressure (EP) additives pour in depressants, corrosion inhibitors, and antioxidants, introducing a unique service when the loads that come into contact are heavy. In industrial grease production, it is important to control the crystallization of the soap so as to maintain a fibrous structure that will prevent the escape of lubricant and allow it to flow at the required rate.

For normal working conditions, a good mineral oil with oxidants and corrosive inhibitors is recommended; for application at high temperatures, one of the synthetic oils may be essential.

It should be noted that the chosen oil should have sufficient high viscosity at the working temperature to maintain an effective oil film on the contact areas (raceways).

Most grease thickening agents are soaps made from sodium, lithium, and calcium. The greases with lithium as a thickener and petroleum oil are most widely applied for antifriction bearings; additives (EP type, antioxidant and corrosion brake additives) are usually provided.

When low friction is required, molybdenum disulfide or graphite may be added to silicone-based greases and complex synthetic greases can be gainfully used for high temperature applications.

The choice of grease type depends on the working temperature: up to 40°C, as in the glands of rotors and combing rolls, a calcium-based grease is recommended; up to 90°C, a sodium-based grease is required; up to 100°C, a lithium-based grease is preferable; and up to 200°C, a silicone-based grease can be applied.

The technique of using grease as a lubricant has not been fully illustrated. However, Qasim (1976) stated that the fibers of soap under the motion action of the contact elements (rolling) are scattered and broken down and the oil provided as a fluid lubricant keeps elastohydrodynamic lubrication between the contact surface (rolling elements) and the raceways and a hydrodynamic lubrication between the cage and its contacting surfaces.

The *viscosity* is a very important characteristic of the lubricant. There are different definitions of lubricant viscosity:

1. *Absolute viscosity (dynamic viscosity)* is measured in poise where 1 poise = 1 dyne–s/cm². Practically all lubricating fluids have less than 1 poise, so it is possible to register the absolute viscosity Z in terms of centipoises (cP), that is, one hundredth of a poise.

2. In the English unit system, the absolute viscosity is converted from centipoises into μ (reyns) where

$$\mu\left(\text{reyns}\right) = \frac{Z\left(\text{centipoise}\right)}{6.9\left(E+6\right)} \tag{A7.1}$$

3. *Saybolt universal viscosity* (SUS) in seconds and *kinematic viscosity* are commonly used in lubrication:

$$Z = \left(0.22\tau - \frac{180}{\tau}\right) \tag{A7.2}$$

where:
Z = absolute viscosity in centipoises
τ = SUS in seconds

$$\text{Kinematic viscosity} = \frac{Z}{P} \tag{A7.3}$$

where *P* is the density of the lubricant in grams per cubic centimeter (g/cm³); for practical application, the average *P* for oils is 0.9 g/cm³.

The most important viscosity grading for practical applications is the one developed by the American Society of Automotive Engineers (SAE).

The relationship between the absolute viscosity Z (centipoises) and typical SAE number oils and temperatures is shown in Figure A7.1.

The viscosity is usually measured in seconds for a certain quantity to be passed via a viscometer such as a Saybolt. These seconds can be approximated to kinematic viscosity with the use of certain techniques. The experiments must be carried out under certain pressure and temperature conditions.

The selection of a lubricant for a certain purpose is based on several factors including temperature, friction, viscosity, volatility, cost, compatibility, flammability, and resistance to atmospheric corrosion. The best lubricant is the one that is the simplest and cheapest but can satisfy functional requirements.

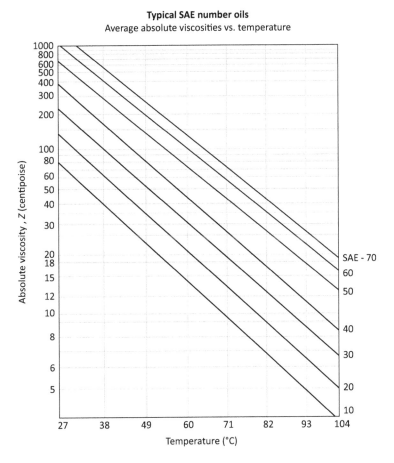

FIGURE A7.1
Viscosity (in Z, SAE) versus temperature. SAE, Society of Automotive Engineers. (From Hall A., Holowenko A.R., and Laughlin H.G., 1961, *Machine Design, Shaum's Series*, McGraw-Hill, New York.)

Usually, 90% of applied lubricants are extended (derived) from crude petroleum in the form of mineral oils and greases. A small quantity of mineral oil will not be satisfactory when the working life is too long or when the heat is too great because wear flakes will be a problem. When sealing is a problem as on a roller or ball bearing, grease should be used and preferred over oils. The simplicity of grease lubrication will give relatively good results unless very high velocities and temperatures make oil a necessity. Also, in some cases, the requirements of low flammability and carbonization will make the use of synthetic oils essential.

Solid lubricants can be subjected to extremely high temperatures and pressures, and despite this they can sustain long trouble-free performance without the need for much maintenance.

The choice of oil lubricant depends on the viscosity grade and the embodied additives. In addition, the most suitable oil for certain cases will be the one with the lowest viscosity that will carry the working load at the required speed.

The use of oil that is too thick will create heat and reduce power loss to overcome viscous friction while a thin oil film may be insufficient to give fluid friction and may cause excessive wear and flakes (debris).

SKF (1991) has tabulated the viscosity classes and conversion tables as shown in Table A7.1.

In addition, according to SKF (1991), there are two types of rotor spindle lubrications:

a. Grease: For a lubricated rotor spindle, the interval of greasing is 5000–8000 working hours (h) for a rotor speed of 40–80 kRPM; grease II is used at 0.20–1.0 g per ball row. A grease gun is used for application.

b. Oil: For a lubricated rotor spindle, the lubricating interval is 2000–6000 h for rotor speeds of 40–80 kRPM; the oil type is Isoflex PDP 65 with a quantity of 0.2–1 cm³ per ball row. The oil is inserted by an oil pistol with a nozzle. In both types of lubrication, the gland of the ball bearing of the spindle may have one or two holes, depending on the rotor model.

For the contact roll assembly of SKF types for a rotor spinning frame, the lubricating interval in operating hours (h) is 12,000/h or 15,000/h, that is, from 2 to 2.5 working years; the maximum speed in revolutions per minute (RPM) is 10,000–16,000; and the lubricant is SKF grease or any good lithium-based rolling bearing grease that has the following properties:

- Working penetration = 265–295 mm/10 at 25°C
- Working temperature range = −30°C to 120°C
- The required amount per bearing is 0.8–1.70 g

TABLE A7.1

Viscosity Classes and Viscosity Conversion Table

ISO Viscosity Grade		Mean Viscosity at 40.0°C mm²/s (cSt)	Limits of Kinematic Viscosity (cSt) at 40.0°C mm²/s	
			Min.	Max.
VG	10	10	9.0	11.0
VG	22	22	19.8	24.2
VG	46	46	41.4	50.6
VG	68	68	61.2	74.8
VG	100	100	90.0	110.0

ISO Viscosity Grade		Saybolt Universal Sec. (SUS)	Redwood °F	Sec. (RHH)	°F
VG	10	55–67	100	74–90	70
		44–54	130	37–45	140
VG	22	104–127	100	191–233	70
		64–79	130	48–59	140
VG	46	214–261	100	493–603	70
		104–127	130	74–90	140
VG	68	318–388	100	731–893	70
		152–186	130	106–130	140
VG	100	469–573	100	1179–1429	70
		210–256	130	140–171	140

Note: Under the international SI system, kinematic viscosity is expressed in meter square per second (m²/s). The relationship between this unit and the figures given in centistokes (cSt) is 10^{-6} m²/s = 1 mm²/s j = 1 cSt. Thus, the mm²/s column in the table corresponds to the centistoke values (cSt).

Source: SKF Textilmaschinen-Komponenten GmbH, 1991, *SKF Almanac*, 7th revised edition, SKF Textilmaschinen-Komponenten, Stuttgart, Germany.

In addition, good calcium-based rolling bearing greases are applicable if they have the following properties:

- Working penetration = 220–250 mm/10 at 25°C
- Drop point = 100°C and service temperature range = −30°C to +60°C
- 0.5 g lubricant per bearing side

SKF complex soap grease has the following properties:

- Working penetration = 220–250 mm/10 at 25°C
- Drop point = 100°C and service temperature is −30°C to 60°C

General Notes

1. The oiling team in the spinning mill must follow the instruction of the machine maker for lubrication, quantity, quality, type, and so on.

2. A high-grade lithium-based rolling bearing grease must have the following properties:

- Working penetration $= 265\text{–}295$ mm/10 at 25°C
- Drop point $= 190°C$ and temperature range $= -30°C$ to 120°C

A lithium-based rolling bearing grease with properties as listed previously can be used or a complex soap grease.

It is well known that the play area of the lubricants is generally when they are inserted or supplied between mating surfaces to minimize wear and friction and in some cases to carry away the generated heat. Therefore, in the following paragraph, we will write about friction inside the bearing, either journal or antifriction, with different types of bearing (ball, roller, needle, etc.), whether radial or thrust. The friction in the bearing plays two roles: heat generation (coefficient of friction f) and power loss due to the resistance of motion. Referring to the coefficient of friction, especially in journal bearings, there are several forms of calculation such as the Petroff equation, McKee equation, and Somerfield number charts.

The Petroff equation is

$$f = 2\pi^2 \left[\frac{\mu N'}{p}\right] \cdot \frac{D}{C} \qquad (A7.4)$$

where:
  f = Coefficient of friction
  $\pi = 3.14 \left(\dfrac{22}{7}\right)$

  μ = Lubricant viscosity in reyns
  N' = Speed of spindle (shaft) in revolutions per second (RPS)
  p = Bearing pressure based on projected area

$$= \frac{\text{Dynamic reaction}}{\text{Bearing length } L \times \text{journal diameter } D}$$

$$= \frac{R}{L \times D}$$

where:
  D = Journal diameter
  R = Diameter clearance between journal and bearing (journal is the part of the spindle end that is dwelling inside the journal)
  L = Length of bearing

The Petroff equation assumes that there is no radial load (no dynamic reaction) and no leakage. It is an approximation for a light loaded bearing.

The McKee equation is as follows:

$$f = 473 \times (E-10)\left(\frac{zN}{p}\right) \cdot \frac{D}{C} + k \tag{A7.5}$$

where:

f, D, C, p = As defined before

z = Absolute viscosity of lubricant at its running temperature in centipoise

N = Spindle (journal) speed in RPM

k = Constant = 0.002 for $\dfrac{L}{D} = 0.75 - 2.80$

$\left(\dfrac{zN}{p}\right)$ = Bearing modulus

Figure A7.2 shows the graphical relationship between the coefficient of friction f and the bearing modulus (ZN/P). It is clear that the slope and intercept of the straight-line portion in the thick film region depends on the $\dfrac{C}{D}$ and $\dfrac{L}{D}$ ratios.

The Somerfield number is a dimensionless value used in calculating the coefficient of friction and in designing the journal bearing.

Legend for Figure A7.2

f = Coefficient of friction
Z = Absolute viscosity in cP at running temperature
N = Journal (shaft) speed in RPM
P = Pressure on bearing $(= R_d/L \times w)$

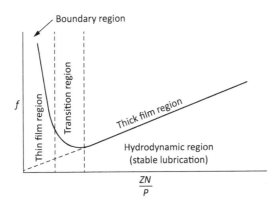

FIGURE A7.2
Coefficient of friction (f) in bearing modulus (ZN/P). (From Hall A., Holowenko A.R., and Laughlin H.G., 1961, *Machine Design, Shaum's Series*, McGraw-Hill, New York.)

R_d = Dynamic reaction
L = Bearing length
w = Bearing width

For the antifriction bearing, the following rules of the coefficient of friction are suggested by Rengasamy (2002):

f = 0.0015 for a single-row radial ball bearing

f = 0.0011 for a cylindrical roller bearing (radial load)

f = 0.0013 for a thrust ball bearing (thrust load)

f = 0.0025 for a needle bearing

f = 0.0010 for a self-aligning radial bearing

Roughly, we can estimate that the average value of f for a journal bearing is obtainable by multiplying any previous values of f by 100.

As the coefficients of friction are calculated, we must then consider heat generation and power loss due to the friction resistance of the bearing.

The friction force created inside the bearing during its operation results in heat generation, which increases the bearing temperature and decreases the lubricant's viscosity.

As the heat builds up, the oil can leak out and if a dry friction condition is attained, this may lead to the bearings seizing.

Therefore, in designing the bearing, the value of heat dissipation must be greater than the heat generated, as is the case in the rotor of the rotor spinning machine where the base is finned (spiral channel) to increase the surface area of heat dissipation.

The maximum allowable temperature increase is generally 60°C. The following formula is used to calculate the generated heat H_g:

$$H_g = f \times L \times D \times P \times V \ (\text{N. m}/\text{s or J}/\text{S or W}) \tag{A7.6}$$

where:

H_g = Heat generated in N.m/s, J.m/s, or watts (W)
f = Coefficient of friction
L, D = Length and diameter of journal in meters

P = Pressure on the journal due to dynamic reaction $P = \dfrac{R_d}{L \times D} \ \text{N/m}^2$
R_d = Dynamic reaction in the bearing
V = Surface speed of journal in meters per second

The heat dissipated must be greater than the heat generated, otherwise external cooling is needed (water or fan).

After heat, we must study the power loss due to friction in the bearing.

$$\text{The torque (resisting)} = f \times R_d \times r \ \text{N/m} \tag{A7.7}$$

where:
   f = Coefficient of friction
   R_d = Dynamic reaction in newtons
   r = Journal radius in meters
   T = Resisting torque in N·m

The power loss in resisting friction is

$$P = T \times \omega \qquad (A7.8)$$

Summary Points

1. The lubricant is mainly applied to minimize friction between two moving parts. In addition, the following relative merits are important:

   a. Lubricants play the role of seals to stop impurities from being fed between surfaces.

   b. To minimize the generated heat, absorption and transfer are used.

   c. Lubricants decrease the wear of moving bodies.

   d. Lubricants facilitate load transmission from one surface to the other.

   e. Lubricants protect the mating surface from corrosion.

2. The different groups of lubricants are

   a. Air (gas, etc.)

   b. Oils (liquid)

   c. Solids (molybdenum disulfide and graphite)

   d. Semi-liquids (greases)

3. Most lubricants (liquids) are derived from crude petroleum (mineral oils). Oils from animals, synthetics, and vegetables can also be applied.

4. Oils have many advantages as they can be applied and circulated easily, chemical additions can be incorporated, and finally they can be recovered and refreshed.

5. Lubricant viscosity can be expressed as

   a. Viscosity class to ISO VG 10–VG 100.

   b. Centistokes (cSt) that range from 10–100 at 40°C.

   c. Limits (range) of kinematic viscosity at 40°C, mm^2/s (cSt): the minimum range is 9–90 and the maximum range is 11–110.

    d. The Saybolt universal second (SUS) that ranges from 61–218 as average values.

    e. Redwood °F ranges from 100 to 130 in two classes only.

    f. Sec. (RHH), which ranges from 82 to 41, 212 to 53.5, 548 to 82, and 1304 to 155.5 as average values against only two classes of °F: 70 or 140.

6. Centistokes CST $= (10^{-6} \, m^2/s = 1 \, mm^2/s)$.

7. EP = Extreme pressure additives incorporated into the bearing oil.

8. The majority of lubricants used in rotor spinning mills are mineral oils and grease.

9. Absolute viscosity Z in SI units is called *dynamic viscosity* and is measured in Ns/m².

10. Kinematic viscosity γ = dynamic viscosity η divided by ρ, that is, $\gamma = \zeta/\rho$, ζ in Ns/m² and ρ in kg/m³.

11. The heat dissipated from the bearing can be calculated by (British units):

$$H_d = \frac{(DT + 33)^2}{k} \times L.D \text{ ft-lb / min} \tag{A7.9}$$

where:

H_d = Heat dissipated in ft-lb/min

L, D = Length and diameter of bearing in inches

k = Constant = 31 for a well-ventilated bearing or = 55 for an air cooling bearing

T = $T_b - T_a$, the difference between bearing surface temperature T_b and the temperature of the air T_a around the bearing housing in °F

But we can write

$$T = (T_b - T_a) = \tfrac{1}{2}(T_b - T_a) \tag{A7.10}$$

where T is the operating oil temperature in °F (1 ft-lb/min = 0.0226 W).

Review Questions

Q#1: Define each of the following:

    a. Absolute viscosity (dynamic viscosity)

    b. Saybolt universal viscosity in seconds

    c. Centipoises and reyns

    d. Kinematic viscosity

Q#2: What are the two types of rotor spindle lubrication?

Q#3: Write Petroff's equation for calculating the coefficient of friction f.

Q#4: What is meant by the bearing modulus?

Q#5: How can the generated heat during bearing operation be calculated?

Q#6: How do you calculate the pressure P inside the bearing?

Q#7: Write a formula for calculating the resisting torque on the journal. What is the value of the power loss?

Q#8: What are the relative merits of the different lubricants?

Answers to Review Questions

Q#1:

a. Absolute viscosity (dynamic viscosity) is when a force of 1 dyne is required to keep a movable plate at a speed of 1 cm/s; when the area of the plate is 1 cm² and is separated from a stationary plate by an oil film that is 1 mm thick, then the oil will have a viscosity of 1 poise.

b. The Saybolt universal viscosity in seconds (SUS) is dependent on the absolute viscosity Z in centipoises (cP). See Equation A7.2.

c. The reyn is a measure of viscosity in the British system and it depends on the absolute viscosity Z. See Equation A7.1. A centipoise is one hundredth of a poise for practical applications.

d. Kinematic viscosity in centistokes (cSt) is dependent on absolute (dynamic) viscosity Z. See Equation A7.3.

Q#2: The two types of rotor spindle lubrication are

a. Grease with a greasing interval of 5000–8000 working hours for a rotor speed of 40–80 kRPM; the quantity of grease is 0.2–1.0 g per row. A grease gun is used for application.

b. Oil with a lubricating interval of 2000–6000 h for a rotor speed of 40–80 kRPM. The oil type is Isoflex PDP 65 (SKF distributer) with a quantity of 0.2 cm³ required per ball row. The oil is inserted using an oil pistol with a nozzle.

In both types of lubrication, the gland of the rotor spindle bearing may have one or two holes depending on the rotor model.

Q#3: See Equation A7.4.

Q#4: Bearing modulus $= ZN/P$.

where:

Z = Absolute (dynamic) viscosity in (cP)

N = Journal RPM

$P =$ Pressure inside the bearing $= R_d/L \times w$

$R_d =$ Dynamic reaction in newtons

$L, w =$ Length and width of bearing in meters.

The coefficient of friction and the design of the sleeve bearing are bearing modulus dependent.

Q#5: See Equation A7.5.

Q#6: Bearing pressure P depends on projected area $= R_d/L \times w$. See answer to Q4.

Q#7: See Equation A7.6.

Q#8: See Summary Point 1.

Review Problems

Q#1: For a rotor of a rotor spinning machine, an oil is used to lubricate the spindle bearing. When the working temperature is 40°C and the viscosity is 46 cSt, what is the absolute (dynamic) viscosity?

Q#2: For Problem 1, calculate the kinematic viscosity. Take $\rho = 0.9$ g/cm³.

Q#3: Calculate the viscosity in reyns for Problem 1.

Answers to Review Problems

Q#1: See Equation A7.2.

$$Z = \left(0.22\tau - \frac{180}{\tau}\right)$$

$$= \left(0.22 \times 46 - \frac{180}{46}\right)$$

$$= (10.12 - 3.91)$$

$$= 6.21 \, \text{cP}$$

Q#2: See Equation A7.3.

Kinematic viscosity $= Z / P$

$$= 10 / 0.9$$

$$= 11.10 \, \text{cP}$$

Q#3: See Equation A7.1.

$$\therefore \mu = \frac{10}{6.9 \times 10^6}$$

$$= 1.4493 \times (E-6)\mu$$

Bibliography

Elhawary I.A., 2014, Rotor dynamics, Lecture notes, Alexandria University, Alexandria, Egypt.

References

Hall A., Holowenko A.R., and Laughlin H.G., 1961, *Machine Design, Shaum's Series*, New York, McGraw-Hill.

Qasim S.H., 1976, *Machine Design*, Basra University Publisher, Basra, Iraq.

Rengasamy R.S., 2002, *Mechanics of Spinning Machines*, NECUTE, Delhi, India.

SKF Textilmaschinen-Komponenten GmbH, 1991, *SKF Almanac*, 7th revised edition, SKF Textilmaschinen-Komponenten, Stuttgart, Germany.

Appendix VIII: Systems of Maintenance

Different systems of maintenance can be applied in the spinning industry (short- or long-staple types) with different varieties of spinning techniques for rotor spinning machines. These maintenance systems are

1. Work-to-failure maintenance
2. Time-based preventative maintenance
3. Conditioned maintenance

Each of these maintenance systems will be explained briefly as follows:

1. *Work-to-failure maintenance*: In the cotton or wool industries, many traditional machines have duplicate core processes and the spinning machines are usually run until they fail or break down. Loss of production is not effective and spare parts can be used during the normally brief repair period. There are small advantages to recognizing a machine's expected failure time, so vibration measurements are used to check the repair quality.

   The latest versions of the spinning machine (rotor spinning machine) are high-tech, either in their design or in their technological features. And with the possibility of applying Six Sigma for product quality in rotor spinning plants, such high-value products lead to the tendency to sometimes let even large unduplicated machines operate to failure. As such, it is worth knowing what is going wrong in advance so that a final failure can be predicted. This information may be obtained by using mechanical vibration spectrum analysis indications from standard measurements; however, the repeated damage resulting from these failures mostly increases both the repair cost and production loss during the extended shutdown period. With the use of conditioned monitoring methods applied in mechanical vibration measurements for maintenance, as explained in Broch (1980), these extra costs can be greatly reduced while the time span between shutdowns can be maintained at its maximum value.

2. *Time-based preventative maintenance*: It has been found that for many machines, if they run for another period of the same length without any maintenance after a breakdown, then the number of machines that break down in the second period will be no more or no less than in the first period compared to when maintenance had been carried out between the intervals (periods) (Broch, 1980). Therefore, there is

now a marked tendency to replace the fixed servicing or renovation periods of machines with fixed periods of measurement of each individual machine's condition. And it is only when the measurements indicate that a certain machine needs it that repairs are made.

3. *Conditioned maintenance*: When the mechanical vibration measurements and their analysis (also referred to as condition monitoring) are done correctly, they will not only help in determining the situation of a specific machine but also in allowing (via tracking trends for individual elements in the spectrum) the prediction of when such individual elements are likely to reach unacceptable levels.

This is termed *predictive maintenance* and permits the long-term planning of work to be carried out. For example, it helps the engineer as a decision maker in purchasing the essential spare parts in advance, so that the storage of large spare parts in stock is avoided. Also, the maintenance team can be practically trained for the repair type that will come up, so a minimum amount of time is consumed with a maximum level of reliability. The use of condition maintenance will minimize maintenance costs and unscheduled breakdowns.

When mechanical vibration detections are embedded into a maintenance program, it is common to use a maintenance team to operate that program. Two types of personnel are usually employed in the team. The first type is the technician who will carry out the actual measurements via a specified technique to register the vibration signal and direct it to the maintenance engineer (second member of personnel in the team) for later analysis and evaluation. The engineer must collect technical data such as rotor speed, bearing geometry, number of rolling elements, and so on, to help him or her later diagnose any detected faults. Engineers will also compile available data and information on acceptable vibration limits from machine makers and other organizations.

Looking at the instruments for on-condition maintenance (OCM) in use today, we find that the simplest technique is to apply a vibration meter to perform a simple, complete mechanical vibration reading in a single frequency band. Most of the standardized vibration detections lie in this class. The reading may allow the investigation of major faults that actually develop, but does not permit diagnosis or reliable prediction.

In 1978, it was found in the United Kingdom that the application of OCM in the British industry led to a minimum saving of £180 million with operation plus investigation costs predicted to be £30 million, ensuring a net profit of £150 million, representing a 500% return, or 30% of the total yearly investment in Britain for machines and equipment. This saving arises from the higher availability factor of the textile machines and reduced production losses (Broch, 1980).

ENTEK IRD (1994) has reported that implementing an effective maintenance program based on vibration measurements analysis and correction takes considerable planning and effort. In spite of this, experience has shown that time and effort to implement a program will return back in terms of reducing costs of maintenance and minimizing production loss. Reduced insurance premiums, a reduction in standby instruments, lower energy costs, and enhanced safety are extra side benefits that will make the effort even more worthwhile.

The establishment of a maintenance program has the following steps:

a. Getting management's commitment.
b. Personnel training.
c. Choosing the rotor spinning machines to be included in the program.
d. Grouping the machines in an organized manner according to predictive maintenance software; machines are usually divided into specified groups. Database setting involves detection locations, measurement directions, amplitude units, FFT parameters, overall alarm levels, and spectra band alarms.
e. Establishing the check periods.
f. Establishing data baselines.

Summary Points

1. There are three types of maintenance system:
   a. Work (run)-to-failure (break) maintenance (WFM or RBM)
   b. Time-based preventative maintenance (TBPI)
   c. On-condition maintenance (OCM)
2. The maintenance (team) staff includes a technician and an engineer. The technician's job is data acquisition while the detection engineer is a data analyzer who investigates the faults.
3. The instruments for on-condition maintenance can be divided into three classes. Class 3 is the simplest where a vibration meter is applied in a single frequency band. Class 2 uses two levels of instrumentation. Class 1 performs full analysis and comparison with reference spectra at each moment.

4. The cost-effectiveness of maintenance provides savings that arise from the higher availability of the machines and reduced production losses. Roughly, the savings correspond to 30% of the total annual investment in the United Kingdom for instruments and machines.

5. Vibration measurements must be applied in a maintenance program as they will introduce advantages or cost savings. The maintenance engineer must build his or her decision (decree) on the basis of a machine's suitability for maintenance, the availability of the team (staff), and the optimization of the cost with respect to equipment to achieve economic effectiveness and use the measurements to reduce maintenance costs to improve the factory's financial performance.

6. The use of mechanical vibration measurements for maintenance is dependent on the maintenance engineer's decisions (decrees), which rely mainly on the economy of the process.

7. The new worldwide tendency in industrial maintenance for cotton and wool spinning generally depends on vibration measurements and analysis.

8. Developing an effective predictive maintenance program based on vibration measurement analysis and correction takes considerable planning effort. However, the benefits of program implementation are great, such as the reduction of maintenance costs, the reduction of production losses, lower energy costs, and enhanced safety.

9. Database setting involves (1) detection locations; (2) measurement directions; (3) amplitude units; (4) FFT parameters; (5) overall alarm levels; and (6) spectra band alarms.

In the rotor spinning machine, these techniques are still not applied (Elhawary, 2017). Perhaps in the future, when the technology of the machine's building architecture and design improves as well as the quality of rotor spun through the application of Six Sigma rules, these advanced programs of maintenance may be more widely applied.

Review Questions

Q#1: What are the three systems of maintenance?

Q#2: Define the maintenance staff (team).

Q#3: Mention the different classes of equipment used in OCM.

Q#4: List the different benefits of the implementation of a predictive maintenance program in a rotor spinning machine's plant mill.

Q#5: What is the meaning of database setting in a predictive maintenance program?

Q#6: What is the worldwide tendency in industrial maintenance?

Q#7: What is the main effective factor in the application of vibration measurements and analysis in a maintenance program?

Answers to Review Questions

Q#1: See Summary Point #1.

Q#2: See Summary Point #2.

Q#3: See Summary Point #3.

Q#4: See Summary Point #8.

Q#5: See Summary Point #9.

Q#6: See Summary Point #7.

Q#7: See Summary Point #6.

References

Broch J.T., 1980, *Mechanical Vibrations and Shock Measurements*, Brüel and Kjær, Naerum, Denmark.

Elhawary I.A., 2017, *Mechanics of Rotor spinning Machines*, Forthcoming, CRC Press, Boca Raton, FL.

ENTEK IRD, 1994, *Introduction to Vibration Technology*, Mechanalysis Inc., Columbus, OH.

Appendix IX: Spike Energy

ENTEK IRD (1994) reported that the mechanical vibrations resulting from rolling element bearings and gear problems will normally be in the form of short duration pulses. Figure A9.1 shows the type of vibration generated by a bearing defect such as a flaw on the raceway. The resultant vibration and energy generated only pass over the defective zone (location), resulting in a short duration spike. Despite the true peak amplitudes created being mostly significant, the average or root mean square (RMS) value of the spike pulses is typically very small when a comparison is made with other sources of vibration such as misalignment and unbalance. This is shown in Figure A9.2, where a rectangle surrounds one cycle interval of the spike-pulse frequency. It should be clear that the area of the rectangle "occupied" by the spike pulse (i.e., its RMS value) is very small compared to the peak value. Practice has shown in many cases that rolling element bearings, even in advanced stages of failure, may give little if any significant increase in displacement, velocity, or acceleration measurements.

Based on the previous discussion, it is apparent that measurements of displacement, velocity, and even acceleration are not well-suited for discovering (detecting) rolling element bearing deterioration and other causes of spike-pulse signals. The basic problem is the requirements for RMS responding circuitry combined with other inherent sources of vibration such as misalignment and unbalance that overshadow or dominate the bearing vibration. However, the resolution is simple as an effort has been made to design a true peak-to-peak circuit to detect a response instead of RMS detection via electronic procedures to filter out the other component of inherent frequencies of vibration that have the tendency to mask the bearing vibration. The result of this work is called *spike energy* (SE) (ENTEK IRD, 1994).

As mentioned previously, when the flaws appear in a bearing, the resulting vibration will be in the form of a series of short duration spikes or pulses as shown in Figure A9.1.

Figure A9.1 shows a short-term (40 ms) waveform that was produced on a ball bearing when a small nick was purposefully ground on the bearing's inner raceway. We can see that the pulse interval lasted only a few microseconds. Usually, if the period of a mechanical vibration signal is known, the frequency of the mechanical vibration can be detected by simply taking the reciprocal of the interval; for example, if it takes $(3600)^{-1}$ minutes to complete one cycle of a mechanical vibration, then the frequency of vibration is 3600 cycles per minute (CPM).

In the generated pulses of the defective bearings, as the pulse periods are so short, the interval's reciprocals (frequencies) are typically too high for explanation; a micro flaw is generally defined as a defect that is so small that

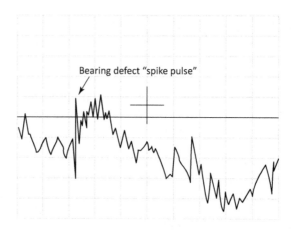

FIGURE A9.1
Short duration spike pulse caused by a small nick ground on the inner raceway of a ball bearing. (From ENTEK IRD, 1994, *Introduction to Vibration Technology*, Mechanalysis Inc., Columbus, OH.)

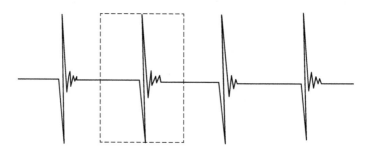

FIGURE A9.2
Spike-pulse vibration caused by problems such as bearing and gear defects may have very low RMS values. (From ENTEK IRD, 1994, *Introduction to Vibration Technology*, Mechanalysis Inc., Columbus, OH.)

it is impossible to see with the naked eye. The pulses created by a micro flaw are typically less than 10 ms. By considering the reciprocal of a 10 ms pulse, the fundamental frequency becomes 100 kHz or 6 kCPM (6 million revolutions per minute [RPM]) as the flaw in the bearing deteriorates. The next phase is a macro flaw, one that is detectable by or visible to the naked eye. Hence, the macro flaw is larger and the interval duration of the created pulse is longer, thus the fundamental pulse frequency is lower. Typically, a macro flaw will create a pulse with an interval (period) exceeding 20 ms, resulting in a fundamental pulse frequency of 50 kHZ (3 kCPM or 3 million RPM) or less as the defects of the bearing continue to build in size and the resultant pulse periods become even longer, resulting in a decrease in fundamental pulse frequency to nearly 5 kHZ (300 kCPM) when the deterioration of the bearing has reached a severe level.

The fundamental features of the SE approach developed by IRD Mechanalysis (ENTEK IRD, 1994) is summarized as follows:

1. As the defective bearing creates vibrations with very high frequencies, use an accelerometer transducer.

2. Filter out high frequencies of 50 kHZ and those below 5 kHZ. This will lead to the detection of the micro flaw defects as SE measurements will significantly increase and a visual inspection of the bearing will give confirmation of a visible flaw. For most predictive maintenance programs, discovering and detecting micro flaws is of great concern as the deterioration of the micro flaw phase can take several months. In addition, eliminating the lower frequency of 5 kHZ filters out most other inherent vibration sources including misalignment, unbalance, electrical frequencies, and so on, which have the tendency to hide (dominate) the vibrations from the defective bearing.

3. Incorporate a true peak-to-peak circuit instead of an RMS detector circuit.

    Establishing the severity criteria chart for gSE is not easy, the difficulty being the returns to the extremely high ultrasonic frequency included. It is well that ultrasonic frequencies of vibration cannot travel well and are attenuated effectively with distance. The ultrasonic frequencies can be easily reflected by the interface between the accelerometer transducer and machine surface, interfaces such as bearing split lines and gasketed surfaces. In such cases, it is vital that the SE reading be taken directly on the bearing housing (cap) and attention must also be paid to how the accelerometer is mounted on the bearing cap (housing). There are three different techniques for mounting the accelerometer: stud or adhesive mounting (best results), drilling or rapping each bearing (most predictive maintenance programs), or a magnetic holder and an extension probe on the accelerometer.

    Figure A9.3 shows the spike energy (gSE) severity chart (ENTEK IRD, 1994).

An additional source of SE is where insufficient lubrication of the bearing results in dry rolling friction. The normal practice is to grease the antifriction bearing when a high SE reading (measured in gSE) is noted to determine if inadequate lubrication is the source of the problem. Despite this, even if the bearing is actually defective, the grease addition will most likely result in a significant decrease of SE. The addition of lubricant simply provides a temporary damping of the intensity of the defect and does not establish the cause. Therefore, if there is a noticeable decrease of the SE after grease addition, ensure that insufficient lubrication is the problem and that the bearing's

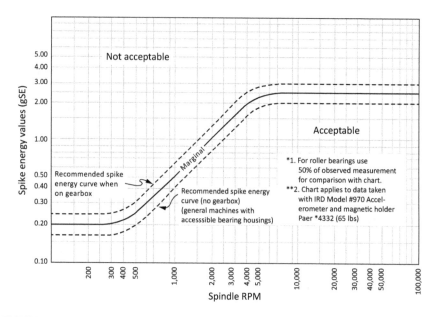

FIGURE A9.3

Spike energy (gSE) severity chart guidelines for ball bearings. (From ENTEK IRD, 1994, *Introduction to Vibration Technology*, Mechanalysis Inc., Columbus, OH.)

mechanical condition is safe. Check the SE several hours or days after lubricant addition to see if the reading values are still low. If they are, then the problem is insufficient lubrication. However, if the bearing is truly defective, the SE levels will mostly return to their original readings (levels). From our point of view, the bearing of the combing roller can be highly defective for several reasons as indicated in the work of Peter R. Lord.

In the gear train of the rotor spinning machine, the gear tooth is impacted by worn, mismatched, or misaligned gears. SE measurements give a much more sensitive indication of developing gear-related problems than velocity or acceleration RMS measurements because gear problems are of an impact-based nature.

When the tangential belt that drives the combing rolls or the rotors has a chunk, it will cause a successive impact on the wharves of the combing rolls and the rotor. This will generate SE; therefore, do not condemn a bearing or gear failure based on SE alone.

Summary Points

1. SE is generated by frequent (repetitive) transient impacts. Such impacts typically occur as a result of surface flaws in rolling element bearings or gear teeth.

2. The accelerometer signal is processed by unique filtering and detection circuitry to produce a signal "figure of merit" connected to the intensity of the original impacts. The figure of merit is expressed in gSE units.

3. The background vibration from background sources such as transportation or running machines inside the mill can damage spare parts or standby instruments. Even new bearings stored in a vibration environment can be damaged by background vibration.

4. Rubbing between rotating and stationary parts (seal rubs, a rotating shaft against a coupling grand, etc.) and loose mechanical and rattling parts can be detected by SE because all of these are generators of SE. Some rotors of the rotor spinning machine can be covered by hand. If such a cover is not well fitted, it can generate SE. Maybe the beater (combining roller) spindle can generate impact actions, then an SE can be developed.

5. A chunk on the tangential belt drive for rotors and combing rolls is a generator of SE.

6. The gears of the gearing diagram of the rotor spinning machine are a source of SE due to mismatching between gears and misalignment; worn teeth in matting is also a source of SE.

7. Any loose and rattling parts produce SE.

8. In a rotor spinning machine, the sources of SE may be
   - Flaws in the spindle (inner race) or in the gland (outer race) that have consequential impacts on rolling elements (balls)
   - The mismatching (mismeshing) of gears in the machine gear due to the misalignment of gear teeth
   - Chunks on the tangential belt drive of the combing rolls and rotors that have successive and frequent impacts on the wharves (driving rolls) of the spindle
   - The driving elements in the rotor machine driving head such as belts, pulleys, and so on

9. The widespread use of the SE technique is still (for textile machines including the rotor spinning machine) too far away, but the future may hide a lot of surprises.

Review Questions

Q#1: What is meant by SE?

Q#2: Explain the IRD Mechanalysis technique for SE detection.

Q#3: Is it necessary to use SE measurements in a new rotor spinning machine?

Q#4: What are the different engineering sources for SE in the rotor spinning machine?

Q#5: If the fundamental frequency of the pulses generated by a micro flaw is 100 kHZ, what is the CPM?

Q#6: Experimentation has revealed that by the time the fundamental frequency has reduced to $\cong 5$ kHZ, bearing deterioration has reached severe limits. What is the value of such a frequency in RPM?

Answers to Review Questions

Q#1: It is ultrasonic waves that are produced due to flaws in the races of a defective bearing; when the rolling element of the bearing impacts these micro flaws, these waves are created.

IRD Mechanalysis has its own technique to detect SE. See Summary Point #2.

Q#2: See Summary Point #2.

Q#3: New (updated) rotor spinning machines use new technology in design and machine construction, which means that the machines are very expensive and any breakdown during running periods will lead to production losses and a loss in capital investment. The detection of SE will help in eliminating defective antifriction bearings, mismeshing gears, and so on. It will also provide reduced noise levels, heat generation, and loss of power requirements. Therefore, the SE is vital for new rotor spinning machines, despite the cost of repair.

Q#4: See Summary Point #8.

Q#5:
$$f_0 = 100 \, \text{kHz CPS} \left(\text{cycles per second}\right)$$

$$\therefore f_1 = 100 \times 1,000 \times 60$$

$$= 6,000,000$$

$$= 6 \, \text{kCPM}$$

$$= 6 \, \text{MCPM} \left(\text{mega cycle per min}\right)$$

Q#6:
$$f_0 = 5 \, \text{kHZ}$$

$$\therefore f_1 = 5 \times 1,000$$

$$= 5,000 \, \text{CPS}$$

$$= 5,000 \times 60$$

$$= 300,000 = 300 \, \text{kRPM}$$

References

ENTEK IRD, 1994, *Introduction to Vibration Technology*, Mechanalysis Inc., Columbus, OH.

Appendix X: G1

A10 G1

Conversion Tables (Broch, 1980)

These tables are: conversion of length, conversion of velocity, conversion of acceleration, conversion of force, conversion of pressure, conversion of work, energy and heat, conversion of power. Then temperature conversion. Finally frequency equation of single-degree-of-freedom system (Free vibrations – transient vibration).

For the relationships between acceleration, velocity, and displacement for sinusoidal vibration, Appendix 10 (A10) has been divided into three groups of tables and figures (i.e., A10 G1, A10 G2, and A10 G3) where G means group, where,

a. A10 G1: Deals with conversion tables (Broch, 1980).

b. A10 G2: Deals with mass moments of inertias & volumes for homogeneous rigid bodies (Bedford and Fowler, 2008).

c. A10 G3: Deals with conversion factors for British and SI units (Bedford and Fowler, 2008).

1. Conversion of Length

m	cm	mm	ft	in
1	100	1000	3.281	39.37
0.01	1	10	0.0328	0.3937
0.001	0.1	1	0.00328	0.03937
0.3048	30.48	304.8	1	12
0.0254	2.54	25.4	0.0833	1

2. Conversion of Velocity

m/s	km/h	ft/min	mph
1	3.6	196.65	2.2369
0.2778	1	54.68	0.6214
5.08×10^{-3}	1.829×10^{-2}	1	1.136×10^{-2}
0.4470	1.6093	88	1

3. Conversion of Acceleration

g	m/s^2	cm/s^2	ft/s^2	in/s^2
1	9.81	981	32.2	386
0.102	1	100	3.281	39.37
0.00102	0.01	1	0.0328	0.3937
0.03109	0.3048	30.48	1	12
0.00259	0.0254	2.54	0.0833	1

4. Conversion of Force

N	kp	lb-ft/s^2 (pdl)
1	0.102	7.2329
9.807	1	71.0
0.1379	1.405×10^{-2}	1

5. Conversion of Pressure

Pa	mbar	mm H$_2$O	atm	in wG	lbf/in^2
1	10^{-2}	0.102	9.869×10^{-6}	4.02×10^{-3}	1.4504×10^{-4}
100	1	10.917	9.869×10^{-4}	4.02×10^{-3}	1.4504×10^{-2}
9.807	9.807×10^{-2}	1	9.678×10^{-5}	0.402	1.4223×10^{-3}
1.013×10^{-5}	1013	$10,332 \times 10^{-5}$	1	3.937×10^{-5}	14.696
249.10	2.491	25.4	2.453×10^{-3}	406.77	3.505×10^{-2}
6908.9	69.089	704.49	6.805×10^{-2}	27.736	1

6. Conversion of Work, Energy, and Heat

J = W.s	kWh	kpm	kcal	Btu	ft·lbf
1	2.778×10^{-7}	0.1020	2.39×10^{-4}	9.48×10^{-4}	0.7376
3.6×10^{-6}	1	3.6710×10^{-5}	860	3413	2.655×10^{-6}
9.807	2.7241×10^{-6}	1	2.3423×10^{-3}	9.2949×10^{-3}	7.233
4187	1.163×10^{-3}	427	1	3.9685	3087.4
1.055	2.93×10^{-4}	107,59	0.25198	1	777.97
1.3558	3.766×10^{-7}	0.1383	3.239×10^{-4}	1.285×10^{-3}	1

7. Conversion of Power

kW	kpm/s	hk	kcal/h	ft·lbf/s	hp
1	102	1.36	860	738	1.34
9.81×10^{-3}	1	1.3×10^{-2}	8.44	7.23	1.32×10^{-2}
0.735	75	1	632	542	0.986
1.16×10^{-3}	0.119	1.58×10^{-3}	1	0.858	1.56×10^{-3}
1.36	0.138	1.84×10^{-3}	1.17	1	1.82×10^{-3}
0.745	76	1.014	642	550	1
2.93×10^{-4}	2.99×10^{-2}	3.99×10^{-4}	0.252	0.216	3.93×10^{-4}
3.52	35.9	0.479	3024	259	0.471

Temperature:

$$F = \frac{9}{5}c + 32 \qquad c = \frac{5}{9}(F - 32)$$

Single-Degree-of-Freedom System

$$m = \text{Mass}(\text{kg})$$

$$k = \text{Stiffness}(\text{N}/\text{m})$$

Resonant linear frequency

$$f_0 = \frac{1}{2\pi}\sqrt{\frac{k}{m}} = \frac{1}{2\pi}\sqrt{\frac{g}{st}} \qquad (\text{A10.1})$$

Resonant circular $= w_a = 2f_0$

$$\text{Frequency} = \sqrt{\frac{k}{m}} \qquad (\text{A10.2})$$

where:
  g = Gravitational acceleration
  st = Static deflection of the mass

For Single Frequency (Sinusoidal) Vibration

Acceleration	Velocity	Displacement
$\partial_0 \cos 2\pi ft$	$\dfrac{1}{2\pi f} \partial_0 \sin 2\pi ft$	$-\dfrac{1}{4\pi^2 f^2} \partial_0 \cos 2\pi ft$
$-2\pi f v_0 \sin 2\pi ft$	$v_0 \cos 2\pi ft$	$\dfrac{1}{2\pi f} v_0 \sin 2\pi ft$
$-4\pi^2 f^2 d_0 \cos 2\pi ft$	$-2\pi f d_0 \sin 2\pi ft$	$d_0 \cos 2\pi ft$

Summary Points

1. The conversion tables are a helpful tool when dealing with units from either the SI or British unit system and can help students, engineers, and readers navigate easily between both types of units.

2. It is advisable for any user of the conversion tables to always start from the cell of unity "1" to move from one system to another fluently.

Review Problems A10 G1

Q#1: A full journal (sleeve) bearing generates 18,500 ft-lb/min (using the McKee equation for f). Find the value of the heat generated in ft-lb/s, kpm/s, hk, kcal/h, and hp.

Q#2: If a group of rotor spinning machines produce 50 tex yarn and consume 2,179,300 kWh per year, what is the value of kWh/rotor? The yarn production per rotor = 170.8 g/h. Assume that the working hours per year are 6000 and the number of rotors/machines is 200.

Q#3: Transfer the kWh/rotor value in Problem 2 to kpm, kcal, Btu, and ft·lbf.

Q#4: In Problem 2, what is the required kWh/kg of yarn production?

Q#5: The suction (negative pressure) inside the rotor of a rotor spinning machine must be no less than 73 mbar (average value) to transport open fibers through the transport channel to the sliding wall of the rotor. What is the value of this negative pressure in Pa, mm H_2O, atm, wG, and lbf/in²?

Q#6: The dynamic reaction in the bearing near the rotor of a rotor spinning machine is 1.5 kp (kilopond = kilogram force) at a rotor RPM of 24,000 and with a residual unbalance of 0.45 g/cm.

Transfer the dynamic reaction value to newtons (N) and lb-ft/s?

Q#7: For a modern rotor of a rotor spinning machine, the maximum allowable rotor circumferential speed (limiting speed) is 210 m/s. What is the value of this speed in km/h, ft/min, and mph?

Q#8: When a rotor starts running, it takes about 10 s to reach its full speed of 170 m/s. What is the value of its acceleration? Convert the value of acceleration to g, cm/s², and ft/s².

Q#9: If the amplitude of forced damped vibrations of a rotor in a rotor spinning machine is 0.20 mm, find the values of the amplitude in m, cm, ft, and in.

Answers to Review Problems AX G1

Q#1: By using Conversion Table 7, the heat generated $H_g = 18,500$ ft-lb/min.

$$H_g = \frac{18,500}{60} = 308.3 \, \text{ft-lb/s}$$

$$= \frac{308.3}{780} = 0.4178$$

$$= 0.4178 \times 10^2$$

$$= 42.6 \, \text{kpm/s}$$

$$= 1.36 \times 0.4178$$

$$= 0.5682 \, \text{hk} \, (\text{metric horse power})$$

$$= 359.3 \, \text{kcal/h}$$

$$= 0.4178 \times 1.34$$

Q#2: $$\text{kWh / rotor} = \frac{2,179,300}{200 \times 6,000}$$

$$= 1.816 \, \text{kWh}$$

Q#3: By using Conversion Table 6, we find

$$\text{kWh/rotor} = 1.816$$

$$= 1.816 \times 3.6710 \times 10^5$$

$$= 6.667 \times (E+5) \, \text{kpm}$$

$$= 1.816 \times 860$$

$$= 1561.8 \, \text{kcal}$$

$$= 1.816 \times 3413$$

$$= 6198 \, \text{Btu}$$

$$= 1.816 \times 2.655 \times 10^6$$

$$= 4.82(E+6) \, \text{ft} \cdot \text{lbf}$$

Q#4: The rotor requires 1.816 kWh, that is, 1.816 kW to work one hour.

For one hour, the rotor produces 170.8 g, that is, 0.1708 kg. Then, the required energy in kWh/kg = 1.816/0.1708 = 10.63 kWh/kg of yarn. This value is sometimes named the specific energy of the rotor to produce 1 kg of rotor-spun yarn.

Q#5: Referring to Conversion Table 5,

$$\text{One amber} = 100 \, \text{Pa} \, (\text{pascal})$$

$$\therefore 73 \, \text{amber} = 73 \times 100$$

$$= 7300 \, \text{Pa}$$

$$= 73 \times 10,917$$

$$= 796.941 \, \text{mm} \, H_2O$$

$$= 73 \times 9,869 \times 10^{-4}$$

$$= 72.0 \, \text{atm}$$

$$= 73 \times 0.402$$

$$= 29.3 \, \text{wG}$$

$$= 73 \times 1.4504 \times 10^{-2}$$

$$= 10,588 \, \text{lbf/in}^2$$

Q#6: Returning to Conversion Table 4, then

$$R_d = 1.5 \times 9.807$$

$$= 14.7\,N$$

$$= 1.5 \times 7.2329$$

$$= 10.85\,lb\text{-}ft/s^2$$

Q#7: Referring to Conversion Table 2, then

$$210\,m/s = 210 \times 3.6$$

$$= 756\,km/h$$

$$= 210 \times 196.15$$

$$= 41.192\,ft/min$$

$$= 210 \times 2.2369$$

$$= 470.0\,mph$$

Q#8: Returning to Conversion Table 3,

$$\text{Acceleration} = \frac{2010 - 0}{10}$$

$$= 21\,m/s^2$$

$$= 31 \times 0.102$$

$$= 2.142\,g$$

$$= 21 \times 100$$

$$= 2100\,cm/s^2$$

$$= 21 \times 32.2$$

$$= 676.2\,ft/s^2$$

$$= 21 \times 386$$

$$= 8106\,in/s^2$$

Q#9: Referring to Conversion Table 1, then

$$0.2\,\text{mm} = 0.2 \times 0.1 = 0.02\,\text{cm}$$

$$= 0.002\,\text{m}$$

$$= 0.2 \times 0.00328$$

$$= 6.56\,(E-4)\,\text{ft}$$

$$= 0.2 \times 0.03937$$

$$= 7.871\,(E-3)\,\text{in}$$

Appendix X: G2

Summary Points A10 G2

1. When the homogeneous rigid body has zero thickness (thin), one of the two masses of inertia I_0 and I_{qc}—polar and equatorial—disappears.
2. According to Case 1, the central equatorial mass moment of inertia is always $I_{qc} = \frac{1}{2}I_0$.
3. When the rigid homogeneous body has a thickness that is not zero, the relations of both I_0 and I_{qc} with regard to the massive geometrical characteristics of the rigid body will be very different.

Review Problems for Figure A10.1

Q#1: If the rotor of a rotor spinning machine has a maximum diameter of 48 mm and a mass of 85.9 g, calculate the polar mass moment of inertia I_0 and the equatorial mass moment of inertia I_{qc}. Treat the rotor as a thin disc.

Q#2: The wharve of the rotor spindle has a length of 36 mm and a diameter of 18 mm. Treat the wharve as a solid cylinder with a mass of 42.0 g and calculate the polar mass moment of inertia I_0 and the equatorial mass moment of inertia I_q.

Q#3: In the rotor spindle, the ball bearing has a diameter of 6 mm. If the bearing is made from steel with $\rho = 7850$ kg/m³, what is the mass of the ball?

Q#4: In Problem 3, what are the values of $I_{x'_{axis}}$ and $I_{z'_{axis}}$?

Q#5: A combing roller of a rotor spinning machine has a spindle with length = 92 mm and diameter = 14 mm. The spindle machine is made of steel $\rho = 7850$ kg/m³. What is the weight of the spindle?

Answers to Review Problems for Figure A10.1

Q#1: Referring to Figure A10.1, Case 2, the polar mass moment of inertia (I_0) around the axis of rotation (axis z') is $(I_0)I_{z'_{axis}} = \frac{1}{2}mR^2$.

$$I_{z'_{axis}} = \frac{1}{2} \times 0.085 \times 0.024^2$$

$$= 2.48 \times 10^{-5}\,\text{kg/m}^2$$

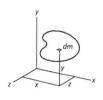

The moments and products of inertia the object in terms of the *xyz* coordinate system are

$$I_{x\,axis} = I_{xx} = \int_m (y^2 + z^2)\, dm,$$

$$I_{y\,axis} = I_{yy} = \int_m (x^2 + z^2)\, dm,$$

$$I_{z\,axis} = I_{zz} = \int_m (x^2 + y^2)\, dm,$$

$$I_{xy} = \int_m xy\, dm,\quad I_{yz} = \int_m yz\, dm,$$

$$I_{zx} = \int_m zx\, dm.$$

1. General situation

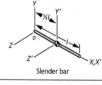

$$I_{x\,axis} = 0,\quad I_{y\,axis} = I_{z\,axis} = \frac{1}{3} ml^2,$$

$$I_{xy} = I_{yz} = I_{zx} = 0.$$

$$I_{x'axis} = 0,\quad I_{y'axis} = I_{z'axis} = \frac{1}{12} ml^2,$$

$$I_{x'y'} = I_{y'z'} = I_{z'x'} = 0.$$

Slender bar

$$I_{x'axis} = I_{y'axis} = \frac{1}{4} mR^2,\quad I_{z'axis} = \frac{1}{2} mR^2,$$

$$I_{x'y'} = I_{y'z'} = I_{z'x'} = 0.$$

2. Thin circular plate

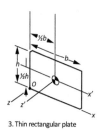

$$I_{x\,axis} = \frac{1}{3} mh^2,\quad I_{y\,axis} = \frac{1}{3} mb^2,\quad I_{z\,axis} = \frac{1}{3} m(b^2 + h^2),$$

$$I_{xy} = \frac{1}{3} mbh,\quad I_{yz} = I_{zx} = 0.$$

$$I_{x'axis} = \frac{1}{3} mh^2,\quad I_{y'axis} = \frac{1}{3} mb^2,\quad I_{z'axis} = \frac{1}{3} m(b^2 + h^2),$$

$$I_{x'y'} = I_{y'z'} = I_{z'x'} = 0.$$

3. Thin rectangular plate

FIGURE A10.1
Homogeneous rigid bodies' mass moments of inertia. (From Bedford A.M. and Fowler W., 2008, *Engineering Mechanics: Statistics*, Pearson, Singapore, London.)

The equatorial mass moment of inertia around the axis perpendicular to the axis of rotation, that is, around axis x' and axis y', is

$$\therefore I_{x'} = I_{y'} = \tfrac{1}{4} \times 0.085 \times 0.024^2$$

$$= 1.224 \times 10^{-5}\, kg/m^2$$

$$= \tfrac{1}{2} I_{z'} = \tfrac{1}{2} I_0$$

Volume = abc

$I_{x'axis} = \dfrac{1}{12}m(a^2 + b^2),$ $I_{y'axis} = \dfrac{1}{12}m(a^2 + b^2),$

$I_{z'axis} = \dfrac{1}{12}m(a^2 + b^2),$ $I_{x'y'} = I_{y'z'} = I_{z'x'} = 0.$

4. Rectangular prism

Volume = $\pi R^2 l$

$I_{xaxis} = I_{yaxis} = m\left(\dfrac{1}{3}l^2 + \dfrac{1}{4}R^2\right),$ $I_{zaxis} = \dfrac{1}{2}mR^2,$

$I_{xy} = I_{yz} = I_{zx} = 0.$

$I_{x'axis} = I_{y'axis} = m\left(\dfrac{1}{12}l^2 + \dfrac{1}{4}R^2\right),$ $I_{z'axis} = \dfrac{1}{2}mR^2,$

$I_{x'y'} = I_{y'z'} = I_{z'x'} = 0.$

5. Circular cylinder

Volume = $\dfrac{1}{3}\pi R^2 h$

$I_{xaxis} = I_{yaxis} = m\left(\dfrac{3}{5}h^2 + \dfrac{3}{20}R^2\right),$ $I_{zaxis} = \dfrac{3}{10}mR^2,$

$I_{xy} = I_{yz} = I_{zx} = 0.$

$I_{x'axis} = I_{y'axis} = m\left(\dfrac{3}{80}h^2 + \dfrac{3}{20}R^2\right),$ $I_{z'axis} = \dfrac{3}{10}mR^2,$

$I_{x'y'} = I_{y'z'} = I_{z'x'} = 0.$

6. Circular cone

Volume = $\dfrac{4}{3}\pi R^3$

$I_{x'axis} = I_{y'axis} = I_{z'axis} = \dfrac{2}{5}mR^2,$

$I_{x'y'} = I_{y'z'} = I_{z'x'} = 0.$

7. Sphere

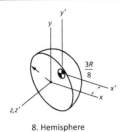

Volume = $\dfrac{2}{3}\pi R^3$

$I_{x'axis} = I_{y'axis} = I_{z'axis} = \dfrac{2}{5}mR^2$

$I_{x'axis} = I_{y'axis} = \dfrac{83}{320}mR^2$ $I_{z'axis} = \dfrac{2}{5}mR^2$

8. Hemisphere

FIGURE A10.1 (CONTINUED)
Homogeneous rigid bodies' mass moments of inertia. (From Bedford A.M. and Fowler W., 2008, *Engineering Mechanics: Statistics*, Pearson, Singapore, London.)

Q#2: Referring to Figure A10.1, Case 5, the polar mass moment of inertia $(I_0) = I_{z'_{axis}} = I_{z_{axis}} = \frac{1}{2}mR^2$

$$\therefore I_0 = \left(I_{z'_{axis}}\right) = \frac{1}{2} \times 0.042 \times 0.009^2$$

$$= 1.701 \times 10^{-6}\,\text{kg/m}^2$$

The equatorial mass moment of inertia is

$$(I_q) = I_{x'_{axis}} = I_{y'_{axis}} = m\left(\frac{1}{12}.\ell^2 + \frac{1}{4}R^2\right)$$

$$= 0.042\left(\frac{1}{12} \times 0.036^2 + \frac{1}{4} \times 0.009^2\right)$$

$$= 0.042\left(1.08 \times 10^{-4} + 2.025 \times 10^{-7}\right)$$

$$= 4.5445 \times 10^{-6}\,\text{kg/m}^2$$

It is important to note that I_q in the previous problems must be the central equatorial mass moment of inertia, and it is then written as I_{qc}.

Q#3: Referring to Figure A10.1, Case 7, then

$$\text{volume} = \frac{4}{3}\pi R^3$$

$$= \frac{4}{3}\pi \times 0.003^2$$

$$= 1.1304 \times 10^{-7}$$

$$\text{mass} = \rho \times \text{volume}$$

$$= 7850 \times 1.1304 \times 10^{-7}$$

$$= 8.87364 \times 10^{-4}\,\text{kg}$$

Q#4:

$$I_{z'_{axis}} = I_{y'_{axis}} = I_{x'_{axis}} = \frac{2}{5}mR^2$$

$$\therefore \text{polar}(I_0) = I_{z'_{axis}}$$

$$= \tfrac{2}{5} \times 8.87 \times 10^{-4} \times 0.03^2$$

$$= 3.19 \times 10^{-7} \text{ kg/m}^2$$

Here, $I_0 = I_{qc} = 3.19 \times 10^{-7}$ kg/m^2.

Q#5: Referring to Figure A10.1, Case 5, then

$$v = \pi R^3 \times \ell = \pi \times 0.007^2 \times 0.092$$

$$= 1.4155 \times 10^{-8} \text{ m}^3$$

$$\text{mass} = v \times \rho$$

$$= 1.4155 \times 10^{-8} \times 7850$$

$$= 0.111 \text{ kg}$$

$$\text{weight} = \text{mass} \times 9.81(g)$$

$$= 1.09 \text{ N (newton)}$$

Appendix X: G3

Summary Points A10 G3

1. Group 3 of Appendix X is very similar to Group 1, but there are some extra unit factors mentioned, such as 1 kip = 1000 lb = 4448 N (force units). Therefore, 1000 lb must be written as 1000 lbf (pound force, not pound mass).

2. Any of the three groups included in Appendix X will be helpful tools for the users of this textbook.

Review Problems for Table A10.1

Q#1: If the acceleration of a rotor of a rotor spinning machine is 21 m/s^2, what is its value in ft/s^2, in/s^2, and g?

Q#2: The rotor-spun yarn cheese has a 3-kg mass. What is its value in slug?

Q#3: In Problem 2, what is the weight of the cheese in N, lbf, and kip?

Answers to Problems for Table A7.1

Q#1:

$$21 \text{m/s}^2 = 21 \times 3.281$$

$$= 68.901 \, \text{ft/s}^2$$

$$= 21 \times 39.37$$

$$= 826.77 \, \text{in/s}^2$$

$$= 21/9.81$$

$$= 2.14 \, \text{g}$$

Q#2:

$$3 \, \text{kg} = 3 \times 0.0685$$

$$= 0.206 \, \text{s/ng}$$

TABLE A10.1

G3 Conversion Factors

Time	Acceleration
1 min = 60 s	$1\text{ m/s}^2 = 3.281\text{ ft/s}^2 = 39.37\text{ in/s}^2$
1 h = 60 min = 3600 s	$1\text{ in/s}^2 = 0.08333\text{ ft/s}^2 = 0.02540\text{ m/s}^2$
1 day = 24 h = 86,400 s	$1\text{ ft/s}^2 = 0.3048\text{ m/s}^2$
	$g = 9.81\text{ m/s}^2 = 32.2\text{ ft/s}^2$

Length

1 m = 3.281 ft = 39.37 in

1 km = 0.6214 min

1 in = 0.08333 ft = 0.02540 m

1 ft = 12 in = 0.3048 m

1 mi = 5280 ft = 1.609 km

1 nautical mile = 1852 m = 6080 ft

Mass

1 kg = 0.0685 slug

1 slug = 14.59 kg

1 t (metric ton) = 907.18 kg = 68.5 slug

Angle

1 rad = $(180/\pi)$ deg = 57.30 deg

1 deg = $\pi/180$ rad = 0.01745 rad

1 revolution = 2π rad = 360°

1 rev/min (RPM) = 'J 047 rad/s

Force

1 N = 0.2248 lb

1 lb = 4.448 N

1 kip = 1000 lb = 4448 N

1 ton = 2000 lb = 8896 N

Area

$1\text{ mm}^2 = 1.550 \times 103\text{ in}^2 = 1.076 \times 10\ 5\text{ fl2}$

$1\text{ m}^2 = 10.76\text{ ft}^2$

$1\text{ in}^2 = 645.2\text{ mm}^2$

$1\text{ ft}^2 = 144\text{ in}^2 = 0.0929\text{ m}^2$

Work and Energy

1 J = 1 Nm = 0.7376 ft-lb

1 ft-lb = 1.356 J

Power

$1\text{ W} = 1\text{ Nm/s} = 0.7376\text{ ft-lb/s} = 1.340 \times 10\sim 3\text{ hp}$

1 ft-lb/s = 1.356 W

1 hp = 550 ft-lb/s = 746 W

Volume

$1\text{ mm} = 6.102 \times 10\text{-}5\text{ in}^3 = 3.531 \times 10\text{-}8\text{ ft3}$

$1\text{ m}^3 = 6.102 \times 104\text{ in}^3 = 35.31\text{ ft}^3$

$1\text{ in}^3 = 1.639 \times 104\text{ mm}^3 = 1.639 \times 10\sim 5\text{m}^3$

$1\text{ ft}^3 = 0.02832\text{ m}^3$

Pressure

$1\text{ Pa} = 1\text{ N/n}^2 = 0.0209\text{ lb/ft}^2 = 1.451 \times 10\text{-}4\text{ lb/in}^2$

1 bar = 105 Pa

$1\text{ lb/in}^2\text{ (psi)} = 144\text{ lb/ft}^2 = 6891\text{ Pa}$

$1\text{ lb/ft}^2 = 6.944 \times 103\text{ lb/in}^2 = 47.85\text{ Pa}$

$1\text{ Pa} = 1.4500\text{ lb/m}^2$

Velocity

1 m/s = 3.281 ft/s = 39.37 in/s

1 km/h = 0.2778 m/s = 0.6214 m/h
 = 0.9113 ft/s

1 mph = (88/60) ft/s = 1.609 km/h
 = 0.4470 m/s

1 knot = 1 nautical mph = 0.5144
 m/s = 1.689 ft/s

Source: Bedford and Fowler (2008).

Q#3:

$$W = 3 \times 9.81$$

$$= 29.43\,\text{N}$$

$$= 29.43 \times 0.2248$$

$$= 6.62\,\text{lbf}$$

$$= 6.62/1000$$

$$= 6.62 \times 10^{-3}\,\text{kip}$$

References

Bedford A.M. and Fowler W., 2008, *Engineering Mechanics: Statistics*, Pearson, Singapore, London.

Broch J.T., 1980, *Mechanical Vibrations and Shock Measurements*, Brüel and Kjær, Naerum, Denmark.

Index